PROLOGUE

I THINK MOST PEOPLE remember 'firsts': their first car, first love, first job. Most firefighters remember their first big fire. I vividly recall mine, back in the days when fires and fire weather were far more predictable than today.

It happened on a hot, dry, windy Saturday in October 1971 following a dry winter. The evening before, 1 October, a bushfire broke out near a rubbish tip at North Turramurra adjoining Ku-ring-gai Chase National Park in Sydney's northern suburbs, making the horizon to the west of our home glow orange as flames lit up the night sky.

Despite being quite close geographically, 10 km to our west as the crow flies, it was about a 40-minute drive to get there by car as you had to skirt around the perimeter of the national park – probably a 15–20 km ride. To reach us the fire would have to jump Cowan Creek and burn through kilometres of bushland. My dad, Jack, a volunteer firefighter, didn't seem too concerned about it.

Most of the local bushfire brigades had been asked to head over to North Turramurra early the following morning to help local volunteers and full-time fire brigade officers to carry out a backburn intended to stop the fire burning downhill to the Bobbin Head

marina, jumping Cowan Creek and then coming into the suburbs
of Duffys Forest and Terrey Hills. Backburning is a firefighting
tactic that involves lighting a new fire ahead of an uncontrolled fire
to remove the fire fuel before the main fire gets there. If it works,
a backburn can stop a fire in its tracks. But they are also risky.
Because they are often carried out as a last resort in weather conditions that cause fires to spread quickly, sometimes backburns escape.
It is always a calculated risk.

On Saturday morning a gusty westerly wind sprung up.
I remember looking west from our front yard and noticing that the
big column of brown smoke was getting bigger and darker. That was
a bad sign. Mum and Dad had explained to me when I was younger
as we looked out over the hills during the long, hot summers when
there was often bushfire smoke in the distance, that black smoke
meant flames were leaping into the tree canopies and burning eucalyptus leaves, or that the air had suddenly become very dry, making
fires burn more intensely.

Once fires started to burn into the crowns of the trees (known
as crowning), burning pieces of bark, twigs and leaves could be
carried up into the smoke column and dumped ahead of the main
fire, starting new ones – spot fires. The sort of vegetation that
grows in Sydney sandstone soils can be very dry and flammable
so spot fires are always a concern. Spot fires can cause the main
fire to spread even faster as the flame front is sucked forward to
meet the new, smaller fires. Dad explained to me how fires draw in
oxygen at the base and how very big fires create their own weather.
He told me about the heatwave and massive fires of January 1939,
when he had seen a smoke column high above a massive bushfire
turn into a storm cloud and then heard thunder, as lightning
flashed out and started new fires miles away. Today we call this a
pyrocumulonimbus cloud or a fire-generated storm, an extreme

FIRESTORM

'Greg Mullins knows what he is talking about when it comes to bushfires and other natural disasters after some 50 years on the front line – he is an acknowledged global expert. He explodes the myths and exposes the ignorance, especially in relation to climate change "supercharging" the increasing intensity and frequency of the fires. Greg's book should be required reading for all Prime Ministers, their governments, and policy authorities. It is no excuse to say "I don't hold a hose, mate"'!

Dr John Hewson AM
Professor, Crawford School of Public Policy at ANU,
and former Liberal Federal Opposition Leader

'Greg Mullins is a born leader. He possesses qualities that do not always distinguish those in public life. He is forthright and competent and he holds a hose. Quite early in his long and exceptional career as a firefighter he observed that bushfire behaviour across eastern Australia was changing. Nobody is better qualified now to talk, in a practical way, about our nation's response to climate change. We will ignore his conclusions, quite literally, at our peril.'

Bob Debus AM
Minister for Environment, New South Wales 1999–2007
Minister for Emergency Services, New South Wales 1995–2003
Attorney General, New South Wales 2000–07
Minister for Home Affairs, Australian Government 2007–10

'Greg Mullins shows us clearly that without urgent and meaningful action to reduce the country's greenhouse gas emissions we have only just seen the beginning of summers from hell. He draws on a lifetime of experience in fighting fire and managing fire services as well as a deep knowledge of the increasing effects of the changing climate on weather conditions that are supercharging bushfires. Greg successfully skewers the myths developed by self-serving parties to minimise the developing catastrophe of a changing climate that is threatening humanity. This book deserves to be taken notice of.'

Naomi Brown
Former Chief Executive Officer, Australasian Fire and
Emergency Service Authorities Council (AFAC)

'The impact of climate change on wildland fires is real, severely increasing their intensity and the damage they cause. This is a global crisis that is no more apparent than what is happening in both California and Australia. *Firestorm* brings to the forefront the environmental and human impacts of these catastrophic wildland fires and the policies that influence them.'

Ken Pimlott
Former Fire Chief, Californian Department of
Forestry and Fire Protection (CAL FIRE)

FIRESTORM

FIRESTORM

GREG MULLINS

VIKING
an imprint of
PENGUIN BOOKS

VIKING

UK | USA | Canada | Ireland | Australia
India | New Zealand | South Africa | China

Viking is part of the Penguin Random House group of companies whose addresses can be
found at global.penguinrandomhouse.com.

Penguin
Random House
Australia

First published by Viking, 2021

Cover photography: scorched earth by Getty/Andrew Merry; firefighter by Getty/Adam
Roper/EyeM; smoke by Getty/Jeff R. Clow; flames by Shutterstock.com/EC Photos.
Cover design by James Rendall © Penguin Random House Australia
Atmospheric stability diagram on page 297 by James Rendall © Penguin Random
House Australia: lightning by Shutterstock.com/klyaksun/Protasov AN; rain by
Shutterstock.com/A_kela; flames by Shutterstock.com/Mikhail Bakunovich/d1sk/
Archiwiz; trees by Shutterstock.com/wichuda suwandee/surassawadee/VikiVector;
smoke by Shutterstock.com/disk.
Typeset in Adobe Garamond by Midland Typesetters, Australia

Every effort has been made to trace creators and copyright holders of quoted
and photographic material included in this book. The publisher welcomes
hearing from anyone not correctly acknowledged.

Printed and bound in Australia by Griffin Press, part of Ovato, an accredited
ISO AS/NZ 14001 Environmental Management Systems printer.

A catalogue record for this
book is available from the
National Library of Australia

NATIONAL
LIBRARY
OF AUSTRALIA

ISBN 978 1 76104 091 7

penguin.com.au

MIX
Paper from
responsible sources
FSC® C009448

For my grandsons Eamon and Oli, and their future safety.
I want you to know that I tried my best.

CONTENTS

event that until recently was quite rare and, back in 1971, almost unheard of.

As I looked at the thickening smoke column, even at twelve years old I could imagine all of the dynamics fuelling the fire, and that it would probably become bigger and ever more ferocious as the day warmed, dried, and the wind strengthened. My brother Terry, a talented competitive cyclist, was away on a long-distance road race, but my sisters Kim and Robin joined me out the front gazing at the smoke with trepidation. We knew that fires could spread very quickly, and the tension in the air was palpable as we looked at the tinder-dry bush surrounding our house. Everybody in the area was on edge, a familiar feeling during the fire season for anyone who lives in the bush.

The phone rang late in the morning, and Mum called out to Dad. It was Terry's best friend, Geoff Carr. He and his teenage sister Jenny were at home alone, their parents away for a couple of days. Geoff could see the huge column of smoke and didn't know what to do. They lived in the last house on Bibbenluke Avenue, a small acreage on the western edge of Duffys Forest adjoining Ku-ring-gai Chase National Park. The house was surrounded by bush at the top of a steep, wooded slope facing west. It would be the first property to be impacted if the fire jumped Cowan Creek.

Dad made another phone call and I remember him looking concerned. He'd found out that the backburn at North Turramurra had escaped, and the fire was expected to jump the creek soon. Local fire trucks were making their slow way back to Duffys Forest, but it would be at least an hour or more before they arrived. Geoff and Jenny were in the direct path of the flames and needed urgent help.

Dad hurried out to our garage and started loading tools into the first new car we had ever owned, a Hillman Hunter station wagon. A rake, a hoe, a shovel, an axe, some old sacks, and a couple of metal watering cans. He put on an old pair of khaki overalls, and heavy boots, and grabbed his ever-present broad-brimmed felt hat. Then he looked over at me.

'Well, hurry up!' he said. 'You'd better get dressed.'

I didn't need any further encouragement. I could see that Mum wasn't impressed, but also that she wasn't going to stop me. I quickly put on long cotton pants, a long-sleeved shirt and my heavy walking boots. Dad handed me an old, cotton broad-brimmed hat and a pair of heavy gardening gloves. I had some experience helping with small hazard reduction burns in our backyard and down the road on my mate Phil Bush's property, so I already knew some of the basics.

It was about a 10-minute drive to Geoff's house, but as we neared the boundary between Terrey Hills and Duffys Forest we saw a small line of stationary vehicles, a police car parked across the road with its single blue light flashing, and a police officer gesturing for everybody to turn around. Dad drove up to the roadblock as everyone else went the other way. The officer waved dismissively at us to go back the way we'd come. We were now much closer to the fire and I could see the huge brown, black and white smoke column twisting and 'boiling' into the sky, the smell of smoke now very strong. It made sense that the road was being closed.

Dad stepped out of the car and got uncharacteristically heated when the young policeman wouldn't listen and said he wouldn't let us through. 'Listen, son, when you've got time, go and read the bloody law,' he said. 'I'm a bushfire brigade officer and you can't stop me. I tell *you* what to do, not the other way around. You better get out of my bloody way because I'm heading for that,' he said, pointing at the clouds of billowing smoke.

Dad had earned the affectionate nickname 'Bloody Jack'; he wasn't one to swear, but he was quite partial to the word 'bloody'. He liked to explain that it wasn't really a swear word, but instead 'the great Australian adjective', citing a poem of that title to prove his point.

I was a bit shocked seeing Dad get angry, because he was always so even-tempered. He jumped back behind the wheel, gunned the engine, and shot past the officer and a couple of other cars. I turned around and saw the officer looking at us with his mouth wide open. Dad by now seemed a bit sheepish and mumbled something like, 'He got my Irish blood up.' There was tacit agreement that I wouldn't tell Mum.

Even though it was the middle of the day, the sky was turning dark as we approached Geoff and Jenny's place a few minutes later. The smoke was starting to blot out the sun, which looked like a huge orange ball in the sky. Trees were bending over in the strong wind, and dust and dry leaves were swirling around in little eddies. Dad had taught me that if it was hot enough and dry enough for the bush to burn, then the wind became the dominant factor in driving the speed and ferocity of a fire. I saw Dad's brow furrow as he looked around and watched the strong wind gusts. This wasn't good.

As we pulled into the driveway, black, grey and white ash was falling all around us. It was stiflingly hot and the smoke was getting thicker. The fire front hadn't arrived yet, so we couldn't see any flames, but it was pretty obvious that it had spotted over Cowan Creek and was now roaring up the steep hillsides towards us, belching huge clouds of writhing smoke.

Dad told me that the main fire front had about 2 km still to cover, but that the westerly wind would push it very quickly up the steep slopes rising from the creek. Fire accelerates uphill (see

Appendix). I first learned this from Dad, and was about to witness it firsthand. Years later I would study the science of it.

The speed of a bushfire can double up a 10-degree slope, and quadruple up a 20-degree one. Radiation, mostly light energy or electromagnetic radiation, is the main way that bushfires transfer heat. Radiation emitted by the flames heats and dries out living and dead vegetation, making it easier to ignite and burn when sparks and flames arrive. When fires burn uphill, convection – the vertical movement of super-heated air – also plays a role. Heat from the flames below preheat the bush above, driving out any moisture and making it catch fire and burn far more quickly and intensely. Without the preheating from below, fires slow down by a similar rate when burning downhill. I should have been horrified, but I was fascinated by the idea that the fire was like a living thing, racing towards us, consuming everything in its path, driven by the elements.

Dad parked the car on the eastern side of the house in a cleared area (which he explained was sheltered as much as possible from the wind, sparks and fire) and started to unload the tools. Geoff and Jenny ran out to help us, looking very relieved. We soaked the hessian sacks in tubs of water and filled containers with water from the nearby dam. Dad climbed a ladder and cleared leaves from the roof gutters, stuffed the drainpipes with wet rags and then filled the gutters with water. He went inside the house and put a ladder up to a hatch in the ceiling, placing a bucket of water in the loft space, ready to douse any small fires in the ceiling cavity.

Then, finally, the fire arrived in all its fury. The sound! It was something I'll never forget – a roaring sound like a jet taking off, together with the strengthening wind whistling and howling through the trees. We had to squint because of the dust, sand and leaves being blown into our faces. Dad yelled at us to grab the wet sacks, watering cans and buckets and to keep an eye out for small fires.

He told us that as the huge flames arrived and the main fire roared through, we should either shelter on the opposite side of the house out of the wind, or go inside if necessary, crouching down low near the back door. If the house caught fire, he said, we should stay inside for as long as we could, stay low, then run outside onto burnt ground when it got too hot, or when the smoke got so thick that we couldn't breathe.

Suddenly I wasn't so sure that I wanted to be here. I started to feel scared and asked Dad how he would stay safe outside. I remember him stopping, looking at me, then stooping down and putting his hands on my shoulders, looking me in the eye.

'It's all right, mate,' he calmly replied. 'I've seen a lot of these. I'll be just outside with the hose. We're all going to be okay.' His calmness was infectious and incredibly reassuring. In my experience, the best fire commanders seem to become calmer as the situation gets more perilous: you can't absorb and process information when panicking, and panic is infectious. Military scholars call this cool, calm effect 'command presence', and it is a key element in effective leadership, but I wouldn't know this for years. Back then, with the fire bearing down on us, it was just Dad's way of doing things. In all my years I never saw Dad get flustered at a fire.

I remember asking him once at a particularly challenging fire how he could stay so calm. He looked at me, shrugged, and said, 'Well, I didn't light the bloody thing. No point getting worried. All we can do is try our best and hope that's enough.'

Soon the main fire front arrived on the opposite side of the narrow road that turned into a fire trail and wound its way down to Cowan Creek. As the inferno raced up the steep slope, we could see flames in the tops of the trees about 100 metres away from us.

The flames were twice as high as the telegraph poles, and the radiant heat coming off the huge flames, even at this distance, was scorching. Black smoke looked like it was boiling up into the air and occasionally the inferno generated fireballs high up in the smoke column as super-heated eucalyptus oil exploded into flames. We had to shout to be heard.

Then came the spot fires. Within moments of the flames cresting the hill we were surrounded by small blazes, dozens of them, as red hot sparks rained down around us. I remember being more fascinated than frightened. Even though the lawn near the house was mown short, the grass was dead and there were lots of dry leaves, pieces of bark and twigs from all the gum trees across it, waiting for an unlucky spark. More than enough fuel under these weather conditions.

As embers landed on the lawn, small fires would start and the wind would immediately whip them into a frenzy, driving them straight towards the house and garden. The flames weren't that high on the lawn, but I watched as big black patches of scorched grass grew as if by magic, fringed by flames that almost came up to my knees. We beat at them with the wet sacks and threw buckets of water, while Dad used a rake and his boots to stamp out flames. A few of the stringy-bark trees caught fire and flames raced up the trunks. This showered us with embers and even more spot fires broke out.

There was a petrol-powered pump at the dam – a luxury in those days of no reticulated or 'town' water. Dad used the hose judiciously to attack the larger spot fires and to wet the western side of the house bearing the brunt of the wind, embers, and raging fire front.

I recall looking to the north and seeing two separate spot fires in the thick bush about 100 metres away with flames starting to roar into the treetops. The fire had jumped the road there too and would soon threaten many other homes to the east, including the home of a school friend, Andy Frodsham. Even the local bushfire brigade station would come under threat. As we raced around trying to put out the spot fires I hoped that Andy and his family were safe, but knew there was no way we could get to them or help them, as we were in the fight of our lives just trying to save Geoff and Jenny's house.

Thankfully the new fires were flanking past us, driven by the wind, not coming at us directly. I saw that the flames were much smaller on the flanks than at the wind-driven head of the fire, and they moved more slowly because they didn't have the wind pushing them forward. Dad shouted that when the tankers arrived, the fire-fighters would try to pinch out the fires by working along the less intense flanks towards the raging head of the fire, then try to knock down the head fire working from behind from burnt ground, which was the safest place to be.

The smoke was incredible. My eyes stung and my throat was raw from trying to breathe the hot, choking fire gases. At one stage Dad saw me struggling to breathe and told me to get down low, because fire draws air in towards the base of the flames, meaning there is usually some clear, cooler air close to the ground. I hit the ground and realised that getting down low also reduced my exposure to radiant heat – a big killer in bushfires, as I subsequently learned when I became a member of the bushfire brigade in 1972.

We continued running from spot fire to spot fire, and I lost track of time given the amount of adrenalin that was surging through me. At some stage night fell and I realised we had been working for hours as the fire jumped over the top of us, then came back on the other three sides as the night wore on. Thankfully, when the fire doubled back it was against the wind and burned with less intensity than when it first roared up the hill towards us.

Later on, Dad asked Geoff and Jenny if they were okay to be left alone. All around us the bush was alive with orange sparks, burning logs and trees – it actually looked quite beautiful, and it was eerily quiet, the wind having dropped just after nightfall. By this time the local fire trucks had returned in force from the other side of the national park. There was a bushfire brigade crew and tanker working near the house, and an urban fire engine in front of my mate Andy's house down the road. Andy's house was safe. It was over, even though the fire was still burning deep in the national park. I helped Dad load up the station wagon, and we headed home.

That was my first big fire, and I was hooked. I had found it frightening, fascinating, and awe-inspiring. I was only twelve, and I often reflect on the fact it was that day I decided that when I grew up, I was going to be a firefighter.

INTRODUCTION

THE 2019–20 AUSTRALIAN BUSHFIRE season, now known as Black Summer, was magnitudes worse than the first fire I fought back in 1971, worse than the previously most devastating New South Wales fire seasons in 1994 and 2013, or anything else we'd ever seen. As bushfires burned from border to border, thankfully sparing metropolitan suburbs of Sydney, Newcastle, the Central Coast and the Illawarra, Sydney was blanketed in smoke and on several days air quality was worse than that of notoriously polluted megacities such as New Delhi or Beijing. In time, the smoke drifted as far south as Melbourne, and grew so thick it circumnavigated the world before it dispersed. As of March 2020, an estimated 24 million hectares had burned nationally, destroying 3,094 homes and about 5,900 other buildings. At least 417 people died due to the effects of smoke, 35 people in fires, and an estimated three billion vertebrate animals were killed or displaced. Some species may have been driven to extinction. Economists believe that more than $100 billion in damage was done to the economy.

Firefighters lost their lives in the line of duty, including three US firefighters aboard a large air tanker that crashed in southern New South Wales; a young father who died when a freak weather event picked up his fire truck and dumped it on its roof; and two other young fathers who died at Bargo near Sydney when a tree crashed through the cabin of their fire truck. I was a couple of kilometres from that scene when it happened, following a long day of fighting fires as a volunteer firefighter, and remember the shock and despair when we received the news. Nothing can prepare you for the news that a colleague has been killed on the job.

At the height of the disaster, the Prime Minister of Australia, Scott Morrison, took a family holiday to Hawaii, returning early when it became clear that public opinion demanded it. When asked to account for himself, he told the media, 'I don't hold a hose, mate, and I don't sit in a control room.'

I do hold a hose and have done since I was twelve years old. I've also sat in many control rooms, although more often these days it's in a command car or a tanker within spitting distance of a raging fire front, back on the front line as a volunteer.

During my 50 years of fighting fires, including my 39-year career with the NSW Fire Brigades, later Fire & Rescue NSW, I specialised in bushfire control and studied bushfire control in France, Spain, Canada and the USA in 1995. I worked through the ranks until 1996 when I was promoted to Assistant Commissioner, at thirty-seven the youngest person ever appointed to that rank, and became Director of State Operations in 2000. In that role I coordinated responses to major bushfires in 2001, 2002 and 2003. Prior to that, the commissioner had detailed me to coordinate operations during major bushfires in 1997, and from that year on I represented him on the peak planning body for bushfires: the NSW Bush Fire Coordinating Committee.

In July 2003, I took over the helm from Vice Admiral Ian MacDougall, who, before becoming Commissioner of the NSW Fire Brigades, had been Chief of the Royal Australian Navy. I was the first person to come through the ranks and be appointed as both Chief Fire Officer and Chief Executive Officer, running both operations and the business side of the brigade, and it gave me the opportunity to steer a path that I had mapped out in earlier roles.

As Commissioner I oversaw many major operations including providing logistical support for medical teams sent to Banda Aceh, Sri Lanka and The Maldives after the 2004 Indian Ocean earthquake and tsunami; coordinating the response to the 2006 Blue Mountains fires; providing support to Victoria following the 2009 Black Saturday bushfire disaster; providing support to Queensland in 2011 following cyclones and floods; sending specialist rescue teams to New Zealand in 2011 following the Christchurch earthquake and to Japan following the 2011 Tohoku earthquake and tsunami; coordinating the response and aftermath of the deadly 2011 Quakers Hill Nursing Home fire; and coordinating the response to the 2013 Blue Mountains, Central Coast and Southern Highlands bushfires.

I travelled the world, was elected as President and Board Chair of the Australasian Fire and Emergency Service Authorities Council (AFAC), the peak council for fire and emergency services in Australia and New Zealand; studied at the US National Fire Academy; addressed fire, police and ambulance chiefs in the UK; and became a Country Director of the International Fire Chiefs' Association of Asia (IFCAA). Amazing opportunities and experiences, but, like everything, it came at a price, and eventually I needed a rest. I loved my time as a career firefighter but decided to retire in January 2017 as the second-longest serving chief in the organisation's 133-year history. An historian told me that the only person who served longer than me, beating me by a year, was found dead at his desk in HQ in 1913. I decided not to try to replicate his length of service and possibly also how it ended. Time for me to leave.

I immediately rejoined the rural fire brigade where I had volunteered and fought many fires beside my dad years before. As a Fire & Rescue NSW superintendent remarked to me when we met at a major bushfire in the north of the state in early 2019, 'When it's in your blood, it's in your blood!'

All this is to say: I know fire. I've been involved in some way in just about every major bushfire event in NSW for decades, and helped to coordinate responses to many interstate bushfire emergencies such as Black Saturday in 2009.

I can tell you with absolute certainty on the basis of my personal observations and on the basis of unchallengeable scientific data and research findings that our climate is changing. Humanity is responsible. If we do not change course urgently, the situation will become worse, with the possibility that human habitation of this planet could in future become untenable as we witness increasingly dramatic climate shifts and extreme weather events. I feel compelled to set out the case for the urgent environmental and political changes needed to preserve the safety of our children and grandchildren. I will explain how we in Australia are now in the midst of a rapidly escalating climate emergency that requires a worldwide mobilisation, similar to wartime or at the very least with the same amount of focus given to the COVID-19 pandemic, if we are to save the planet as we know it and our way of life.

My studies and observations since the NSW bushfire emergency of 1994 led me to the conclusion that we were on an increasingly dangerous trajectory that is now resulting in more frequent natural disasters, increasingly on a scale beyond our ability to manage or control. As temperatures continue to increase, so too does the threat of death and destruction due to out of control fires, heatwaves, storms, cyclones and floods. The 2020 Royal Commission into National Natural Disaster Arrangements, also referred to as the

Bushfires Royal Commission, and independent inquiries into the 2019–20 bushfires in New South Wales, Queensland, Victoria and South Australia, all confirmed and explained how climate change has super-charged bushfire risk. The Royal Commission warned of an increase in 'compounding disasters', something that people in New South Wales endured over the last two years with drought, heatwaves, fires and floods. To top all of that off, we have also endured a worldwide pandemic.

In a country accustomed to periodic major bushfires, Black Summer 2019–20 was far worse than anything we had ever experienced before. Black Summer burned more forest, killed more animals, wiped out more sensitive habitats, destroyed more buildings, and killed more people through the combined effects of fire and smoke than any other fire event in Australian history. Multiple scientific studies have concluded that the weather that drove the fires (see Appendix) could not have happened without the warming effect of climate change, which made the landscape drier and more prone to burning. Other studies have concluded that by 2040 such formerly unprecedented weather conditions will be 'average', and by 2060, our hottest, driest year – 2019 – will be considered 'exceptionally cool'.[1]

The world is rapidly warming, with Australia warming slightly faster than other parts of the world. Weather patterns are changing, and we are increasingly experiencing weather extremes that are driving bigger, more intense fires, storms, floods and other natural disasters. The science is clear – global warming is driven mainly by carbon dioxide emitted from the burning of coal, oil and gas. Human-caused climate change is a well-established fact, not something that is still being evaluated and queried, except by people and organisations spreading misinformation, debunked conspiracy theories and sometimes outright lies.

Scientists mostly use 1870 as a pre-industrial baseline to measure temperature variations reflecting the impact of the Industrial Revolution. Natural systems have been unable to absorb increasing emissions since then, particularly given that another facet of significant population growth has been massive land-clearing and deforestation – removing millions of trees that naturally absorb and store carbon dioxide. Land-clearing continues at an alarming rate in Australia and worldwide.

Despite scientists giving detailed warnings for decades about increasingly calamitous effects, the world continues to fail to rein in ever-increasing greenhouse gas emissions. Australia is a standout because of our lack of ambition and action when compared to most other developed nations. We have one of the highest per capita emission rates in the world, and we are about the fifth largest emitter overall after taking into account the coal and gas that we sell to be burned by other countries. Even when our exports are ignored, Australia ranks in the top 20 emitters globally.

The big-spending 2021 Australian Budget had no new initiatives and minimal spending aimed at reducing our climate risk or increasing the uptake of renewable energy, instead focusing on more fossil fuels, particularly gas.

Statements about 'technology pathways' and the fallacy of 'clean coal' are meaningless and intended to conceal the fact that Australia has no credible plan to reduce emissions, our government instead seeming more focused on the maximisation of profits for the fossil fuel industry for as long possible. In early 2021, Great Britain, the USA and the European Union were considering imposing carbon tariffs on countries such as Australia that have been laggards on climate action. As significant parts of the world retreat from fossil fuels, Australia and our economy will be increasingly squeezed because our government has no strategic plan to replace the revenue, create new industries and jobs, and reposition Australia's economy in a way that proactively reduces greenhouse gas emissions. The political landscape in Australia has seen climate action demonised and weaponised, so the Labor opposition is reluctant to outline any ambitious climate action plans, with internal division about coal mining in particular.

The Bushfires Royal Commission explained that even if governments worldwide start to reduce greenhouse gas emissions immediately, decades of warming leading to more destructive fires and natural disasters still lie ahead of us because of greenhouse gases already emitted. It pointed out that what happens after mid-century will come down to emissions reduction actions we take right now – warming will either start to stabilise and eventually reduce (with immediate strong action), or continue to deteriorate if we stay on the current trajectory.

The best analogy I can think of is a house fire – initially as smoke accumulates it might not be noticed, because hot air rises and it gathers at ceiling height. As the fire increases in intensity and spreads faster, more and more smoke is generated – by the time the smoke reaches floor level, it has driven out oxygen and replaced it with poisonous gases like carbon monoxide, meaning that any occupants will probably die. Even when the fire is extinguished, the toxic gases remain for some time and firefighters cannot enter without self-contained breathing apparatus, often having to use fans to remove the toxic gases.

Greta Thunberg likened the world to a house on fire: the dangerous gases have built up to the point that people are starting to die. Over the last 200 years or so, we haven't noticed the build-up because CO_2 is colourless, odourless, and tasteless. Even if we 'put out the fire' by driving down emissions, it will take a long time to clear out the 'smoke'. We have no time to lose and need to get

serious about saving the Earth as we know it. If the current emissions policy vacuum continues, then our government does not
deserve to stay in office.

I feel a strong sense of duty to try to explain to people exactly
how fire science and practice intersect, how climate change is
placing all of us in increasing danger, and how we will see a worsening of bushfires and extreme weather impacts during our lifetimes.
Black Summer should be a massive wake-up call to all of us, let
alone the many other consequences of a warming climate such as the
slow death and perhaps irreversible destruction of our magnificent
Great Barrier Reef, a natural wonder that my great-grandchildren
may never see. I hope that people who read this book, told through
the eyes of a firefighter who has witnessed climate change driving
worsening fires, realise that the safety of future generations is in
our hands

1

THE GOOD OLD (PREDICTABLE) DAYS

ONE OF MY CLEAREST childhood memories is from when I was about five years old, maybe 1964. I remember waking up with a start; it was probably about 4 am and starting to get light outside as the sun peeped above the horizon (this was years before daylight saving). Still groggy from sleep, it took a few minutes for my mind to clear.

Something didn't feel quite right, and after a while I realised that it was *that* sound. A familiar shudder ran up my spine. Mum and Dad had taught me what it meant. The sound of dry gum leaves brushing against each other as the wind started to gust from the west. I could also hear the sighing of the wind through the needles of the casuarina trees in the bush across the road as the wind strengthened. A sound like no other; subtle, mesmerising, menacing – other kids didn't seem to know what I was talking about when I mentioned it at school. But to me it meant one thing: danger.

Nervously, I sniffed the air coming through the open window. Safe – so far. No smoke. If I smelled smoke, I knew I had to run and wake up Mum and Dad. I tried to get back to sleep, but as I would learn over the next fifty or more years, that probably wasn't going to happen. It was never a good sign when a hot, dry westerly started to blow from the desert before the sun was up, and even worse if it was accompanied by the sweet smell of burning eucalypt leaves.

Over the following years there were many occasions when I would get out of bed, listen, sniff the air, then think back to my childhood. The night before a forecast bad bushfire day always was, and still is, a restless, sleepless one for me and many other firefighters.

Trying to predict how a bushfire season is going to pan out is a combination of art and science. The art being local knowledge, on-the-ground intelligence about vegetation, previous fires and what firefighters call at-risk 'assets' (homes, sheds, livestock, farms etc.), and a general gut feel based on historical experience. The science includes meteorological observations and data, predicted weather patterns, an understanding of climate drivers, and computer sim-ulations predicting whether a particular year is likely to be wetter, drier, hotter, cooler, or simply average.

Back when I started fighting fires with my dad in 1971, and before that, it was more art than science. The science of meteor-ology and the technologies underpinning it compared to today were rudimentary, and the roles of major climate drivers such as the Southern Oscillation Index (to gauge the strength of El Niño and La Niña events), the Indian Ocean Dipole (that also impacts Australia's rainfall and temperature patterns) and the Southern Annular Mode (that influences the position of a belt of westerly winds blowing over Australia) were not well understood (Appendix). Observation and experience played a larger role than scientific projections, as weather forecasting in the 1960s and 70s had difficulty determining what was going to happen in the next few days, let alone in coming weeks and months.

I grew up in a semi-rural bushland environment on the northern outskirts of Sydney, surrounded by beautiful sandstone escarpments, cliffs, deep gullies, white sandy beaches, estuaries and creeks. Pristine native eucalypt forest stretched to the horizon as far as the eye could see and in the 1960s there were many small farms and market gardens.

When my mum, Pat, and dad, Jack, ventured out to Terrey Hills on their old Harley-Davidson motorcycle with the three kids (my brother Terry and sisters Kim and Robin) crammed into the sidecar in 1955, they found dirt roads, a lot of bush and small farms surrounding a 'village centre', namely the post office and shop. My parents had won a building block in a land ballot at a low interest rate, and previously had no idea where Terrey Hills was.

Before they moved there, Mum and Dad first built a garage where they lived while they built the house at night and on weekends. One evening after they moved into the garage, there was a knock on the door. It was Frank Beckman, a local mover and shaker who was on the shire council. Frank explained to them that Terrey Hills was a small community, everybody knew each other, and everybody was expected to do their bit. This might be via the school parents and citizens association, the Red Cross, or the bushfire brigade.

Frank thought that because Dad had been in the air force during World War II and was a builder by trade, he would be a good, practical hand in the local bushfire brigade. From that time Frank became a great family friend and Dad became a volunteer firefighter. He'd already fought fires as a teenager during the January 1939 heatwave, and had been deployed to the Blue Mountains with other soldiers, sailors and airmen during the 1944 bushfires. He'd also fought fires up there in 1952 while helping to build a youth club.

Mum and Dad each had a strong social conscience and were always looking for ways to help others and to do things for the community. They volunteered in community organisations, campaigned against war, and for Indigenous land rights, women's rights and environmental protection. Mum loved the bush and taught me about ecology, the science behind the practice, and how the original custodians of Australia had lived in harmony with the environment and protected it until they were nearly wiped out when our descendants arrived, bringing smallpox and guns. She knew the common and botanical names of all the native trees and wildflowers on our doorstep, and also the invasive weeds that we would pull out on our many bushwalks together with my brother and sisters. She ran a bush regeneration group for years and was trustee of a large bushland reserve that she helped rehabilitate.

Both my parents ended up working in the public service, Mum as a primary school teacher and Dad going on to become a senior works supervisor for the Department of Public Works. He supervised the building of several local schools where Mum later worked. They taught me that public service was an honourable, necessary calling that underpinned a fair and just society, and something to be proud of. They also taught me that you must always be scrupulously honest and have the courage of your convictions, even if it came at personal cost.

I was born in 1959 and as a youngster was always in the bush with my brother and sisters, largely unsupervised by adults: catching yabbies in the creeks, swimming, building cubby houses and tree houses, and exploring the adjacent national park looking for rock engravings and hand stencils left by the area's original inhabitants, the Garigal people. It was great growing up surrounded by beautiful bush: lots of adventures, lots to explore and do, and lots to learn about the native flora and fauna. But part and parcel of that great life was the occasional danger of bushfires.

I was attuned to that danger, perhaps more than other local kids, almost from the day I could walk because of Dad's work as a volunteer firefighter. Fires fascinated me. Living in an area surrounded by dense bushland you had to be aware of the bushfire threat, and my parents made sure that Terry, Kim, Rob and I understood when it was too dangerous to be out in the bush and what to do when fires threatened. They taught us to keep a keen eye on the weather, how dry it was, and particularly on the wind. Hot, dry, windy days were not good days to be out in the bush.

There were many times as we grew up during the 1960s when my brother, sisters and I would overhear conversations about upcoming spring and summer bushfire seasons and what they might bring. There were two revered old-timers, Len Rhodes and Bob Kerswell, who had lived in the area for many years and had always been at the head of local firefighting operations. Len was a virtually toothless World War II veteran and local mechanic who didn't talk much. When he did talk, everybody listened. He always wore big steel-capped boots, khaki shorts and a short-sleeved shirt, even when fighting fires. Bob used to run the local post office and shop and was well dressed and articulate. They were an unlikely pair, but a great team. The local council appointed Bob as honorary (unpaid) Fire Control Officer, and Len as Group Captain, the two most senior officers in charge of the district's volunteer bushfire brigades. My dad and Len got along especially well because they had both been in the air force during the war.

Listening in to their conversations, I started to learn about the wind directions that were dangerous for fires, how heat on its own wasn't necessarily a problem, but when it came on the back of north-westerly or westerly winds with low humidity, then yes, it was a problem. I also learned that rain patterns in preceding months had a lot to do with fire risk, and drought was a major issue because everything would become tinder dry and primed to burn, meaning that on days of hot, dry, windy weather, there could be fire problems.

When it was very dry, fires continued to burn overnight and so-called dry firefighting techniques – using hand tools to cut fire breaks, sometimes combined with small backburns – were less effective because it was impossible to put out every burning log and tree. Sparks could be easily blown by the wind across extinguished fire edges to start new fires.

Most years we would experience the winter westerlies – strong, dry winds at the end of July and mainly in August that would dry out the bush following the regular winter rains that tended to peak in May and June, and sometimes into early July. The winds were cold as well as dry, harsh enough to give me chapped lips and hands that could be very painful.

If the winter rains stopped early, this could spell problems in August because the bush would already be dry by the time the westerlies arrived. If the westerlies were late, however, this created another type of problem, as the strong, dry winds coupled with slightly higher temperatures in September and sometimes through to October markedly increased fire danger, making fires more difficult to contain.

The winter rains were as regular as clockwork. The whole family would usually go camping at Easter and again over the Queen's Birthday weekend in June, and we usually got drenched during both outings. If we stayed at home in June we would enjoy what was known as cracker night, when everybody in the neighbourhood would build bonfires and let off fireworks. Many years it rained, but I noticed as I was growing up that this started to change. Easter became warmer and drier, and rainfall less likely over the June long weekend. The winter westerlies were still strong and dry, but not always freezing cold as they had been in my childhood.

Over the years I recall lots of small fires breaking out when the westerlies arrived – Dad told us how they were hard to extinguish during the day while the wind was blowing, but would then often self-extinguish overnight when it got cold, the wind stopped, and moisture levels increased. They were never that much of a worry and were sometimes left to burn, with Dad saying, 'It's doing a good job clearing out the undergrowth.' If no properties were threatened, Dad and the other veteran firefighters sometimes held off sending in people during the day when the fires were burning more intensely, opting instead to attack them when they died down after dark, the objective being to have the fire 'locked in' (under control) before sunrise the next morning.

Each year Dad noted when certain plants started to flower; when cicadas, Christmas beetles, bogong moths and other insects arrived; and how the weather patterns seemed to be panning out. Mum would watch certain trees closely, such as banksia serrata, and taught us how, during dry periods, their leaves would start to turn yellow, then die, as the tree sacrificed them so that it could survive through a dry period. Casuarinas would be one of the first casualties in a drought – they had relatively shallow root systems and would start to die off when shallow soils dried out. I learned that observing the different trees could sometimes be almost as good an indicator of local soil dryness as the Keetch-Byram Drought Index used by the Bureau of Meteorology.

On walks through the bush on our doorstep, Mum and Dad also showed us how, during very dry periods, eucalypt trees would start to lose bark and leaves in large quantities, increasing fire fuel loads, and when very dry, even sacrifice large branches, which could be dangerous during high winds.

I was fascinated by how Dad had a particular interest in the behaviour of ants. I remember him sometimes saying over the years, even during hot and dry periods, that rain would arrive within the week. He was rarely wrong because he had learned to observe natural indicators, which I'm sure the original inhabitants, who had lived there for tens of thousands of years before us, would have developed to a far more sophisticated level. The tragedy was that we had no way of sourcing the original wisdom and knowledge, because the traditional custodians had been killed by disease or driven out of the area almost two centuries previously.

One of the indicators Dad looked for was swarming ants. His hypothesis was that ants could sense very subtle changes in air pressure if rain was on the way. He reasoned that swarms of workers would seek out more food to store in their nests in anticipation of being prevented from leaving by rain or flooding. He said that if you saw swarms of ants heading for high ground or climbing up walls, then it was 'probably going to piss down!' I never checked this out scientifically, but the indicators Dad used seemed to be pretty much spot on. At least, up until the late 1990s.

In later years he would become puzzled when ants started to swarm but no rains arrived. His take on it back then was that something was changing that was having a subtle effect on weather patterns and air pressure. At the same time, our ability to accurately predict how fire seasons would pan out was reducing. Previously, Dad had always been able to pick the bad fire years.

Back in the 1960s when local fires were burning, my mum, brother, sisters and I would get out hessian sacks, rakes and buckets, ready to protect our home if fires threatened. Sometimes, the day before bad fire weather, Dad would climb a ladder, stuff the downpipes on the roof gutters with rags then fill the gutters with water to stop sparks from setting leaves in the guttering and the ceiling space alight. We only had rainwater tanks in those days, so we didn't have a garden hose, but we knew that our neighbours, the Ham family, had one hooked up to a pump that drew water from a natural spring in the back corner of their block.

Dad was rarely home when fires came close, as he was usually out on a fire truck in charge of a team of local volunteer firefighters. The bushfire brigade shed was only about 200 metres from our house, and I would often lurk outside the door of the radio room listening in to the messages with fascination.

Even better was when my dad told war stories from the fire front. Such as in 1957, the first bad fire season after my parents moved to Terrey Hills, when Dad and his team of volunteer firefighters were cut off by a sudden wind change and had to run through a tunnel of fire to escape, with men on the old fire truck hosing them down as they emerged from the bush, scorched and shaken. Dad was my hero, but very understated, and he never talked himself up – but other people I spoke to said that he had saved the crew that day. I was incredibly proud of him, his practicality, and his quiet bravery. I think I was drawn to firefighting because I wanted to be like him, and I imagined myself in those stories, fighting fires by his side.

Sometimes during the bigger fires that went for days, Mum would go over the road to the Red Cross hut to help make sandwiches and meals to be taken out to the firefighters. My brother Terry, the eldest, would also sometimes go out to fight the flames. Everyone in our small semi-rural community pitched in to do whatever they could, and I couldn't wait to be old enough to jump on the fire truck.

There were countless hot summer nights when Mum would take us out to the front yard and lift me, the youngest, up onto the roof of our old Vauxhall. We'd gaze out over the hills at orange glows from all the bushfires, and wonder which one Dad was fighting.

Mum never said anything, but I knew that she always worried when he was out fighting fires. A couple of times he suffered injuries and had to go to hospital, but it just seemed to be part and parcel of what he did.

If Dad was out at a fire I would often stay awake, listening, and jump out of bed when he came home: blackened, red-eyed, smoky and exhausted. He'd be too tired to answer my tirade of questions about the fire, and would just smile through the soot. He'd have a shower, get a few hours of sleep, go off to work to build someone a house, then study at technical college at night for his clerk of works certificate. Sometimes, he would go out among the flames again when he finally got home the following evening.

I decided way back in the early 1960s what I wanted to do for the rest of my life: fight fires, save lives, do the best I could for the environment, and help people. It was inevitable that I would enter the public service, just like most of our extended family.

The turnout system for the bushfire brigade in the 1960s was not what it is today, with our computer-aided dispatch systems, satellite tracking, and rapid response capabilities with modern trucks, flashing lights and sirens (not to mention brakes that actually work when you hit the pedal!). Like much of life back then, it was a bit more laid-back. In those days, most of the local bushfire brigades were equipped with converted 1940s ex-army Chevrolet and Ford Blitz wagons fitted with tanks, pumps and hoses. They could go anywhere on or off road, but were incredibly slow. There was no protection for the crew, who just held on to the outside of the truck wherever they could. If it got hot you could open the windscreen like a window in a house, and you could take the doors off too – 'air conditioning'.

During the 1968 bushfire emergency I was at school in the after-noon when one chugged past the playground with about fifteen men holding on for grim death. There were no work health and safety rules back then, and no standard uniforms or helmets either. I have a copy of the front page of the *Daily Mirror* newspaper from 23 January 1967 that features a big picture of Dad at a fire direct-ing a team of firefighters dressed in an assortment of overalls, jeans, work clothes, and felt hats. Only two had helmets.

One of the fire shed's most reliable pieces of technical equip-ment was an old blackboard. When a call to a fire was received, if it was assessed as not needing a fast response, which was more often than not given the more laid-back attitude to fire in those days,

A firefighter's protective gear in 1967 stretched to a long-sleeved shirt and felt hat. Dad is the one pointing in this photo of volunteers fighting a bushfire at Belrose. More than fifty years later at Batemans Bay, in my breathing apparatus, radio, helmet and fire resistant hi vis.

someone would chalk a message on the board such as, 'Bushfire at West Head. Have dinner and meet at shed at 7 pm,' then prop it against a tree.

People coming home from work would see the sign as they drove down the main road, and later the old Blitz tanker would slowly drive away with up to a dozen men hanging off its sides. They would come back when the fire was out, however long that took. Sometimes cleaning up the truck afterwards seemed to take an awfully long time, and I remember Mum laughing and speculating once that maybe beer was involved. Still a youngster, I was puzzled because I didn't know how you could use beer for cleaning firefighting equipment.

In about 1969 an old air raid siren was installed on a telegraph pole outside the newly erected brick fire station. The old fire shed had been infested with white ants and was too small to fit the larger tankers that were starting to replace the ex-World War II vehicles. Three blasts on the siren, which could be heard for 5 miles, meant that a tanker crew was required. The sound of the siren sent chills down my spine and was deafening if you were close.

Many times as a youngster I remember the phone ringing, Dad listening intently, nodding and scribbling down notes, wordlessly handing me the key to the siren box with a nod as he finished the phone call, usually holding up three fingers. That meant I had to dash across the road, climb onto the fence, unlock a padlock, and push the button to activate three blasts of the siren. As the siren went through its long wind-down, Dad would jog past in his fire gear, jump on the truck with the other volunteers and head off to the fire. I was so proud of him and he was most definitely my hero.

After fighting my first fire in 1971, I was able to join the Duffys Forest bushfire brigade at age thirteen in 1972, because, as my brother Terry helpfully advised me, 'They're desperate and they'll take *anyone*.'

I wasn't allowed to join Terrey Hills Brigade where Dad had been a deputy captain until I was sixteen, but I had a great time at Duffys, soon becoming an equipment officer. I think just about everyone was designated as an equipment officer, but it made me feel very important. I would walk kilometres to the fire shed every weekend and make sure everything was ready for our infrequent call-outs.

It wasn't long before I was able to get into the thick of things, going to a number of small fires and hazard reduction burns with the other volunteers, and on Christmas Eve 1972 I fought a major fire not far from home that burned about 400 hectares in the national park.

The 1970s were quite busy and I attended many fires, including bushfire emergencies in Hornsby and Baulkham Hills in 1975, the Blue Mountains in December 1977, and then a major fire that destroyed fourteen homes in our local area in December 1979. When I think back to the early 1970s, though, it is a bit of a blur, as it was dominated by the tragedy of losing Terry in a road accident while he was working in Queensland in 1972. His death affected all of us deeply.

In November 1977, in the midst of my Higher School Certificate exams, I had a narrow escape from a fire. It was the day of my biology exam, a scorcher of a day with a Total Fire Ban in place. Trying my best to concentrate, I looked out of the windows of the school hall to see the trees being whipped around by a strong wind as the temperature climbed into the mid-30s, which in those days we considered to be fairly hot.

I finished the exam (which I did very well in!) and ran to catch a bus home. Not long after arriving, the fire siren sounded. I pulled on my overalls and boots, grabbed my helmet and ran across to the fire station. The captain, Ted Carroll, told me that there was a bushfire at Deep Creek, next to Narrabeen Lake. We had a crew of five and headed off with flashing lights and two-tone air horns blaring. When we arrived, it was obvious that the fire had been deliberately lit: there were about five separate ignition points, all of them starting to move quickly as the hot, dry, westerly wind fanned the flames. There was a lot of talk among the firefighters about what they'd like to do to the arsonist.

Ted spoke to the Fire Control Officer, Brian White, who was the council employee in charge of the local bushfire brigades at the time. Brian positioned us on a dirt track where we tried to stop the fire from jumping onto a steep hillside. If the fire got past us, we knew that it would take days to round it up as it would burn into a large area with few access points. The flames were coming through in waves as the strongest wind gusts drove them into the treetops. The fire would then die down again briefly, something Dad had taught me to watch for, as this was when you would try to race in with hoses and douse the flames. Spot fires started to break out on the hillside beside us and things were looking increasingly perilous. We dragged our hoses up the hill to put out spots, but fires started to break out below us as the ember storm intensified. We had to run through the flames at one stage to get to safety and one of the plastic hoses burnt through.

At that stage we realised that we were fighting a losing battle and we would probably have to retreat to safety. There was only one other tanker there, from the Fire Brigade at Narrabeen. Other tankers were coming but still some time away; they were going to be too late. As Ted considered what to do next, I said I would scout down the track a bit to see if the fire had jumped further down – if it had, there was no point staying in the increasingly dangerous spot.

Almost immediately I heard a shout and looked back – flames were towering over the top of the truck, licking one side. Ted was yelling for everyone to get in. As I ran back towards them, flames roared across the track in front of me. Suddenly I was trapped behind a sheet of fire with no way to reach the truck, and I watched as it took off to safety with one hose still dragging behind. I later learned that as everyone had piled into the cabin, in the confusion they'd thought I was with them. Ted felt terrible, but if they'd waited any longer it wouldn't have ended well. He made the right call, as the truck would have caught fire if they'd stayed any longer.

I looked around, fighting the rising panic and trying to work out if I had any options. The hillside was now burning and on the opposite side of the track, the flanks of the fire were racing to join up with the new fires. When this happens, fires 'draw' each other together and create strong air currents and winds. I realised there was no way out – fires on all sides and the strengthening wind was whipping 10-metre flames into a frenzy. I considered running through the flames but remembered the 'rule of thumb', that the depth of flame would be about three times the height. Nobody could survive running through 30 metres of intense flame, and I thought to myself grimly that I would probably be on fire within the first 2 metres, then die a horrible death.

Everything was happening very quickly. I saw a wheel-rut in the track where a truck had been bogged at some time in the past and realised there was nowhere else to go, no refuge, nowhere to run.

I lay facedown in the wheel-rut and tried to scoop some sand across my back. Then the fire came. I felt it suck all the oxygen out of the air around me and I couldn't breathe. When I did get a breath, the air was super-heated and burned the back of my throat. I tried to hold my breath. The heat was incredible, and I could feel it on my back through my overalls. The back of my neck felt like it was burning. I remember feeling light-headed and wondering, 'Is this it?' My immediate thoughts were, 'Mum is going to kill me!' and then, 'Shit! I was going in to apply to join the full-time fire brigade next week – maybe I won't make it.'

I must have passed out, as the next thing I remember was hearing a loud exclamation – 'Ah fuck!' – and a hand on my shoulder. I turned my head to one side and looked up. Col Turnbull, the Deputy Fire Control Officer, was kneeling beside me. Were those tears in his eyes?

'Come on, Scruff, wake up – we've got work to do,' said Col, always the joker. 'No lying down on the job, son!' Col was a former special forces soldier, a former forestry worker, and a renowned fire-fighter. Because I had an afro at the time, he had christened me 'Scruff', a name that caught on.

He walked to his car and grabbed the radio microphone. 'Mobile 26 to Base, I've found him. He's safe and well.' Yep, they were tears, but he was laughing with relief. I'll never forget that.

Col wanted me to go to hospital to get checked out, but I was embarrassed that I'd become separated from the crew and caused them stress. I told everyone I was fine and we kept fighting fires for another few hours.

In that time, the back of my neck blistered up, my face turned bright-red and I developed a splitting headache. When I got home looking like a boiled lobster, Mum was *not* impressed. But I was alive. And I had to study for my maths exam the next day. I never told Mum the full story, but I discussed it with Dad. He agreed that I'd done all the right things. He told me that staying calm and following my training had saved my life.

*

The day that I applied to become a full-time firefighter with the NSW Fire Brigades (I later renamed the organisation Fire & Rescue NSW to better reflect its multiple roles) was memorable. I came home late from my Year 12 formal, and after a few hours' sleep woke suddenly at about 5 am to find Dad shaking me.

'There's a bad car accident at Forest Way and Mona Vale Road, mate – we have to take the tanker up there.' I jumped out of bed, pulled on my overalls and boots, grabbed my helmet, then Dad and I jogged over the road to the fire station. He drove the fire truck and we arrived at the accident within about five minutes. It was a nasty scene. There were two confirmed fatalities, the first that I had been close enough to see, and we provided fire protection for a couple of hours, shielded from the worst of it while police, ambulance and full-time fire brigade officers worked to remove bodies from a crushed car. It was standard procedure for a fire crew to stand by at the scene of a serious accident with hoses because spilled fuels can ignite. I felt really sad for the people who had died. There was a lot of blood in the car which I tried to avoid looking at, as it made me think of how we had lost Terry five years earlier. My thoughts turned to the families who were about to learn that they'd lost family members. To this day, I pity the poor police officers tasked with delivering 'death messages'.

We went back to the station, cleaned the gear, then went home. Dad had a talk to me to make sure I was okay, asking how I felt and what I thought about what we'd witnessed. That helped. Years later I reflected that my lifelong study of martial arts helped me through a lot of situations. The self-discipline and ability to stay calm in threatening situations was a real bonus, particularly back when there was little if any recognition of the mental health impacts on emergency workers and no formal support services available. But I came to realise late in my career that the long-term and cumulative effects of trauma can't be ignored or toughed out.

I learned that brushes with mortality were part and parcel of joining the Fire Brigade. Firefighters (and their families) understand that witnessing death and tragedy is a possibility at every call-out. This unfortunate rite of passage was sobering, but I had another challenge ahead of me. In hindsight it was just what I needed to lighten the mood a little, though I didn't realise it at the time.

My naturally curly hair had grown into an afro by 1977, though not nearly as magnificent as my sister Rob's. Other volunteer firefighters used the nickname assigned to me by Col Turnbull, 'Scruff', and later I would have several others, including 'dunny brush'. My next challenge that morning was to go for my first haircut in about two years, in preparation for my interview at Fire Brigade Headquarters later that morning. I was very apprehensive, probably more about the haircut than the interview.

I headed off to a nearby shopping centre and remember the barber eyeing my frizzy mane warily as I walked in. He asked me what I wanted him to do.

'Make me look like a firefighter!' I said proudly.

I remember him smirking and saying, 'I'm a barber, mate, not a magician.' I loved the poetry of Banjo Paterson and at that moment I'd never had more appreciation for 'The Man From Ironbark' and the bushman terrorised by a shifty barber.

I sat down with great trepidation and the barber took his time throwing a cape over me and laying out his tools of trade: comb, scissors, razor. They looked more like tools of torture to me. I then watched in the mirror with growing dismay as he went to work. It looked like he was trimming a hedge, with great masses of frizzy hair falling to the ground. I saw my ears for the first time in a couple of years and noted to myself, 'Yep. Rob's right. They're big.'

When the barber finished, I asked him for a broom, and, as Mum had made me promise, I scooped up the hair and put it in a pillowcase so that I could take it home to her. The barber thought this was hilarious and let me know it. I didn't know we paid them to editorialise, and decided that I didn't like barbers much, not that I'd met many.

I went home, put on my best pair of flares, headed into Sydney city by bus and train, then walked into Fire Brigade Headquarters in Castlereagh Street. I had no way of knowing it then, but I would later serve from that building for many years.

After sitting down to answer four maths questions and write an essay, I remember as clear as day the recruitment manager, Geoff Hatton, telling me, 'You do realise if we let you in, you'll have to get a haircut, don't you?' Shattered. I already felt like I was bald.

A few weeks later I received a telegram telling me that I had been accepted as a recruit firefighter and had to report for duty at the Fire Brigade Training College at '0800 hours on 31 March 1978'. I was in!

After the initial three-month training course, I was transferred to Number 24 Station, Manly, on Sydney's Northern Beaches, about 25 minutes' drive from home. I was assigned to A Platoon, a shift of six firefighters with a station officer in charge. There were four platoons at each fire station, and we worked a rotating roster: four days on and four days off. Day shifts were ten hours, from 8 am to 6 pm, whereas the nights were 14 hours long, from 6 pm to 8 am the next day. There were set routines for checking trucks and equipment, training periods, study for exams, cleaning the station – and then we'd drop everything when the alarm bells rang and a call-out came over the PA system. We'd race off to all manner of emergencies, then it was back to the station, clean up, replenish, and wait for the next call.

Manly Fire Station had a fire engine that carried all of the hoses and equipment, and had an onboard water tank coupled to a high-capacity pump. We also had a 30-metre ladder truck that was used to rescue people from burning buildings and to project water onto major fires from above.

In those days, structure fires – mainly house and unit fires and the occasional large factory fire – were quite common in the area, so we were busy. We also went to lots of small bushfires, car accidents, and, yes, even cats stuck up trees. I loved it and the crew I worked with was fantastic, helping me to learn the ropes as I studied for mandatory promotional exams that came up regularly.

I later worked as a relieving firefighter, moving between fire stations to cover other people's four-week annual leave breaks, and then for four busy years at Crows Nest Fire Station, responding to incidents from North Sydney to Lane Cove. During this period, I learned a lot about road accident rescue techniques and I also specialised in driving and operating ladder trucks and the large hydraulic platform units commonly known as cherry pickers. From there I went to City Headquarters where I had done the entrance exam back in 1977, the busiest fire station in Australia. I worked up the ranks through the examination and merit system to Station Officer, then District Officer (Inspector) over the course of about a decade, attending many large fires and other incidents along the way.

Just before being transferred to HQ I met my future wife. A friend of mine, Andy, asked me to teach his karate classes while he was away for a couple of weeks. I'll never forget Erris walking into the hall, looking spectacular even in a baggy karate uniform. I was puzzled because she was wearing a brown belt – one belt below black – but I'd never met her before. I found out that she'd been training for years, including at the headquarters of our karate style in Malaysia. To cut a long story short, after getting my breath back after being literally floored by her during a sparring session, I realised that it was love at first kick. It took me over a year to convince her of this though.

In 1984, soon after I transferred to HQ at the rank of Senior Firefighter, one day I received a phone call from an old friend from Crows Nest, Bob Dobson, by then a station officer. He asked me if I would like to relieve at the Bushfire Section where he had just landed a job as Assistant Bushfire Officer. I jumped at the chance.

For the next six or seven years I had the best of both worlds. I would regularly be called in to the Bushfire Section to work on hazard reduction burns in winter, to help train people to use chain-saws and drive four-wheel drive tankers, and to instruct at bushfire fighting courses. In summer I would be called in when there were major bushfires. Before I was promoted to officer rank, I had been to lots of bushfires working as an assistant to incident controllers. There were quite a few senior officers who had only ever worked in the city, so they relied heavily on technical advice from the Bushfire Section when they were commanding bushfire operations. I was very fortunate because most of the senior officers gave me a lot of leeway and listened to my advice, so I was fulfilling roles that other fire-fighters would never have an opportunity to do until they reached officer rank.

In 1989 I was sent to relieve in the vacant position of Blue Mountains Hazard Reduction Officer, working with a small team from Springwood Fire Station to carry out hazard reduction burns Monday to Friday each week. The job involved interacting with local fire brigades, bushfire brigades, and the NSW National Parks and Wildlife Service. I had fought the 1977 fires in the mountains and had a decent working knowledge of the area, one of the most fire-prone in the world, but the 1989 stint gave me a far deeper insight into the unique topography and vegetation that made the area so volatile.

As a station officer from 1990 I spent a lot of time relieving at the Bushfire Section, then in 1991 I was the successful applicant for the position of Sydney Region Hazard Reduction Officer. This position involved responding to summer bushfires to help coordinate operations, conducting bushfire training, and coordinating all hazard reduction work by urban fire crews during winter. I had a Toyota LandCruiser with lights and sirens, a portable weather station, and a small command set-up in the back. I was on call 24/7, but the good thing was that I was no longer on shift work and generally got to spend weekends with the family – our son Phil had arrived in 1985 and daughter Kate in 1987.

It was my dream job; in many ways, something I'd trained for in some capacity my whole life. With the skills, experience and resources of my colleagues in the service, and the expertise and knowledge of experts in the bushfire brigades, forestry and national parks services, we felt at that time that we could adequately handle what Australians thought of as a normal summer.

2

A NORMAL SUMMER

SATURDAY 23 DECEMBER 1990 was shaping up to be hot. Erris suggested that we go to nearby Hornsby to do some shopping, get the kids a treat, and try to escape the heat of the day in the air-conditioning. Phil, then aged five, and Kate, three, were excited about what Santa might bring for Christmas in two days' time. They hoped that one of his helpers might be at the shopping centre so that they could reinforce the fact they'd indeed been good, and also make sure that Santa clearly understood what they were hoping he would bring, as detailed in letters they'd both sent to the North Pole the week before.

It had been a fairly normal Sydney summer so far, but today was going to be hotter than normal. Although only mid-morning, it was already heating up. The humidity was dropping fast, while a stiff north-westerly breeze started to gust and strengthen. A Total Fire Ban had been declared for Sydney, the Blue Mountains, the Illawarra and Hunter regions, with a forecast of Extreme Fire Danger. There had been a few fires in the preceding couple of weeks, and more were expected given today's forecast.

Thankfully though, I thought, any fires today should be 'one-day wonders', because a cool southerly change was forecast to arrive later, which should enable fire services to get the upper hand on any blazes that broke out over the following milder days. Hopefully, firefighters would get to have Christmas Day at home; if not, it wouldn't be the first time that I'd had a Vegemite sandwich and a bottle of water for Christmas lunch somewhere out in the bush.

Theoretically I was off duty, but in my current position relieving for two months at the Bushfire Section as an assistant bushfire officer ('BFO 3'), I was on-call day and night. Erris and I agreed that she and the kids should head to Hornsby in the family car, and I would follow in the fire brigade LandCruiser. There was a chance I would be called in to work, given the way the weather was deteriorating.

Almost as soon as Erris drove off with the kids strapped into their seats in the back of our station wagon, my pager went off. No mobile phones for people of my rank in those days. Just a stark message: 'CONTACT COMMS URGENTLY.'

I switched on the two-way radio and called in. 'BFO 3 Yellow.'

The reply came straight back. 'BFO 3, you are to respond to a bushfire at Killarney Drive, Killarney Heights. On arrival report to D.O. North.'

At least it was in an area that I knew, I thought. As a BFO I was often sent to bushfires in western and south-western Sydney, the Central Coast, and even out to the Blue Mountains. The emergency was closer to home this time, and local knowledge always helped. The role of a BFO was to provide technical advice to the incident controller, liaise with the bushfire brigades, take weather readings, gather intelligence, and help develop the incident action plan. In practice, senior officers often delegated a lot of the operational and command decision-making to the BFOs, as some district officers whose careers had been largely in inner-city areas were far more comfortable dealing with burning buildings than burning bushland.

I pulled out of the driveway, switched on the sirens and flashing red lights, and headed off, a familiar surge of adrenalin making my stomach squirm. It would take me about 30 minutes to get to the fire, depending on traffic, and I was already thinking through what I would need to do when I arrived on scene.

I soon saw our family car up ahead and Erris pulled over with all the other traffic to let me pass. The car windows were open, so I could see the kids' little arms poking out of the back windows waving excitedly as I shot past. I saw the big smiles on their faces. They loved it when I had to head off to a fire. They loved it even more when I went to their kindergarten or school to give a talk to their classes, showing the kids and teachers my firefighting uniform and telling them about fire safety at home and in the bush.

Driving towards Killarney Heights, I thought to myself that at least Erris knew I'd been called in to work this time. How many times had family plans been shattered and Erris and the kids left on their own, or left to wonder where I was when I didn't arrive home from work at the normal time? I have no idea, but to this day she has never, ever complained. Worried, sometimes, I'm sure, but never complained.

She told me once that those times I didn't come home she would turn on ABC radio or the TV news and listen for any bulletins about major fires. If there was a report of a fire, she would put two and two together and assume that I was there and couldn't call. Sometimes I'd get home and the kids would say excitedly that they'd seen me on TV. Maybe they would see me on the news again tonight?

As I drove along Mona Vale Road between St Ives and Terrey Hills, I could see the ominous smoke column in the distance – black and brown and pushing upwards despite the strong wind. Not a good sign. Lots of heat under it driving a swirling convection

column high up into the air. I could picture the flames roaring into
the treetops consuming the eucalyptus oil in the leaves and generating
a storm of embers that would be borne by the wind to start spot fires
ahead of the main fire front. I looked around at the horizon and there
was another big smoke plume south of Sydney. Not my concern, but
I knew that Kevin 'Frosty' Forrest and Graham McCarthy, the two
other BFOs, had probably also been called in to work.

The Killarney Heights fire had started near a picnic area beside
Middle Harbour, raced up a steep hill and then sent burning embers
across six lanes of Warringah Road on the approach to Roseville
Bridge. Flames had then burned up through dry vegetation clinging
to sheer cliff faces and the heavily wooded, steep hillsides above to
immediately impact rows of houses on the ridge tops. Great water
and bush views for the residents, but a nightmare for firefighters
trying to protect them from a raging fire.

Fire engines from the NSW Fire Brigades and bushfire tankers
from the volunteer bushfire brigades (this was before the Rural
Fire Service had been formed, with bushfire brigades administered by local government back then) were racing from house to
house, firefighters rolling out hoses to try to knock down flames and
stop houses from catching alight. It was a suburban area, 'urban /
bushland interface', and strictly speaking the fire was within urban
fire brigade jurisdiction – but both agencies worked together to get
the job done and their respective trucks, skills and equipment were
complementary.

At this point in the operation, and judging from the frantic radio
messages, there was no thought yet being given to how to eventually
control the fire; the entire focus at this stage had to be on saving life
and property. I could hear other urgent radio messages too: out of
control fires on the Central Coast and in the southern and western
suburbs of Sydney. A long, busy day lay ahead.

When I arrived at the fire I found the on-shift district officer, Keith King, working from the back of his station wagon while the mobile command centre, a converted bus, was being set up in a reserve. Keith was a good hand at bushfires and I liked working with him. He was always calm and competent, was always prepared to listen to different ideas, and had a lot of respect for volunteer bushfire fighters. I never quite understood the friction that sometimes emerged between the fire services – it was usually just about egos, rarely about the communities we were both committed to serving and protecting.

He had liaison officers from the bushfire brigade, police and ambulance, and was directing firefighting resources as they arrived. He asked me to be his eyes and ears, which meant driving around the streets, noting down what fire trucks were where, and letting him know my appreciation of the situation and how it might unfold.

I was able to piece together a good 'mud map' of deployments on a sheet of paper and realised there were a few gaps. We needed more fire trucks and firefighters. I started to head back to the control point to brief Keith but an old adage – 'no plan survives contact with the enemy' – soon came into play. Sure enough, it didn't work out as I had planned.

A flank of the fire that couldn't be seen from the roads above had crept around the bottom of a cliff face far below. There was too much smoke to realise what was happening. The flames soon came under the direct influence of the strong wind and within seconds the previously creeping flames were roaring up a small gully that formed a funnel. The sides of the gully closed in on both sides, concentrating heat and flames on a particular point where there were several homes.

I saw the surge of dark smoke and went to investigate. As I pulled up, flames were roaring overhead and spot fires were starting in people's gardens upslope on the other side of the road. I could see the back deck of one of the houses on the lower side was already ablaze and thought I could also make out smoke rising from the roof. The radiant heat was intense, and I tried to stay as low as I could, but that didn't help as the flames were towering above me. It felt like I was caught in a griller and my eyes stung from the acrid smoke.

When I reflected back later, I suspected that the sudden rush of heat and towering flames that poured out of that gully probably unleashed a torrent of burning embers high into the sky, in turn sparking a new fire that I would soon be sent to investigate.

There was only one fire brigade crew on scene. The station officer and three firefighters (the standard crew size) were obviously stretched. They were running full pelt with coiled hoses under their arms, bowling them out and hooking them up to a hydrant so that the high-volume pump on the fire engine would have plenty of water. Normally, being part of the incident management team, I wouldn't get involved in direct firefighting, but I knew that we had a chance of saving the house if we attacked with two lines of hose immediately, rather than just one.

I grabbed a breathing apparatus set from a locker on the side of the fire engine, pulled out two rolls of 38 mm hose and handed one end to the pump operator to connect to the fire engine. I bowled out the hose then dragged it up to the front door, cracking open the nozzle to let air out of the line as the pump operator saw my 'water on' signal. We couldn't hear each other because of the howling wind, and the fire sounded like a jumbo jet coming in to land.

By now, black smoke was pulsing out of roof openings and the eaves, and flames were starting to push through gaps in the roof tiles at the back of the house. After checking that the power was switched off, I kicked in the locked front door and flames immediately curled over the top of me out of the doorway, greedily sucking up air and immediately building in intensity. I squatted down and gave the hallway an initial high-pressure spray, then aimed the nozzle up towards the ceiling where it was hottest. Facedown, a quick blast straight above to dislodge anything about to fall on me, then in I went, staying low where there was still a bit of visibility.

The pumper crew had a second hose and came in from a different door diagonal to me. A golden rule – never attack fires from opposite sides, or smoke, heat, flames and debris can be driven straight onto the other crew. We each worked our way through the black smoke and intense heat into the burning rooms, trying to cut off the spreading fire.

Roof tiles and ceiling plaster were falling in on us and somewhere I heard glass smash. The fire in the roof was a worry – we kept listening for any cracking or creaking sounds, ready to jump under a door frame if the roof caved in, or worse, a hot water tank crashed down. Door frames are reasonably strong and can protect you during a ceiling collapse. We had turned off the power, but we dodged drooping wires, just in case.

I had once been trapped and buried by a collapsing ceiling at a fire in a multi-storey building in North Sydney – on my twentieth birthday. Thankfully I was uninjured, but it was a frightening experience as I lay there, initially pinned facedown by a large beam, mentally checking myself to see if any bones were broken. I wondered for a few minutes if it would be my last birthday, but soon enough other firefighters were calling out to me, digging and pulling away debris, and I was able to crawl out. So I knew from experience that it was best to be wary when inside a burning building. You never knew what could happen in a fire.

After what seemed like half an hour, but was probably more like 10 minutes, we gained the upper hand and confined the fire to the back deck, the kitchen, part of a hallway, and one bedroom. If there had only been one line of hose, the fire probably would have got away and gutted the entire house. It was lucky that I'd been passing just at the right time.

Structurally the house was still sound, so hopefully the owners would be able to have it repaired rather than having to knock it down and rebuild. I knew they would be shattered by what they'd lost, but they were much better off than they might have been. I also knew that it wasn't the building that people worried most about – it was irreplaceable objects such as photos, diaries and the like, which hold special memories. Buildings can be replaced. Loved ones and precious memorabilia cannot.

After making sure there are no people to be rescued, firefighters always try to save and salvage personal effects. This fire was what we called a fairly good save. As I pulled the hose out of the house, I saw that the Christmas tree was still sitting in a corner with presents under it. A bit scorched, but still there. What a Christmas for this poor family, I thought to myself.

I had radioed for help and fire trucks from both the fire brigade and bushfire brigade were on their way from other parts of the fire to help us, because four or five other houses had also been scorched as the fire roared up through the funnel. In the smoke it was hard to tell if any of the others were on fire.

The most common scenario in a bushfire is embers setting fire to leaves in gutters or being blown into roof cavities, setting the ceiling cavity alight, or rear wooden decks catching alight, as in this case. Because of all the smoke from the bushfire, unless you are up close and looking very carefully, sometimes it's almost impossible to tell if the house is alight until the fire is out of control and the building beyond saving.

Media reports regularly quote people saying that they saw houses suddenly 'explode' in a bushfire, which is actually a misnomer. What they are actually seeing is usually the end result of an unobserved ceiling cavity fire caused by embers. At some stage the ceiling collapses and the fire extends into rooms throughout the house, leading to a sudden increase in fire intensity when everything reaches ignition temperature and bursts into flames simultaneously – this is called a flashover. The sudden increase in pressure can blow out windows and force fire gases and flames out of every opening. It is not instantaneous and happens after a fire has been building up for at least 10–15 minutes, often longer. You don't want to be caught inside a room when it flashes over – the cause of many firefighter deaths worldwide.

As I caught my breath outside the now smouldering, steaming house, a bushfire brigade tanker loomed out of the smoke, lights flashing. I saw that the tanker was from Terrey Hills, my old volunteer brigade prior to becoming a career firefighter. I wondered if I knew anyone on the crew. There was someone in particular I was hoping to see.

A familiar voice came from out of the smoke. My dad's! 'Hi, mate. Are you okay?'

'Hi, Dad! Yeah, no worries. Can you guys look after those two houses? I've got two more pumpers on the way and they can deal with the ones on the other side.'

Dad nodded, gave some directions, and volunteers jumped out of the truck and started to pull out hoses.

I walked up and gave my old man a hug. I saw with concern that his eyes were giving him trouble, but when I asked about it, he shrugged it off. 'No, they're fine.'

Years before, he had been at a fire when a spark landed in one of his eyes. He was rushed to Sydney Eye Hospital for treatment and had to wear an eye patch for about a week. Ever since, smoke had badly affected his eyes. Regardless, he still hopped on the fire truck when they needed him. At 66, he was still fit and strong, one of the brigade's most experienced firefighters, and the teacher and mentor of many.

I gave Dad a wave and he gave me a thumbs up. 'See you on Christmas Day!'

It wasn't an uncommon way for us to meet at this time of year, and no doubt we would bore everyone to tears as we compared our fire stories over Christmas lunch.

My radio crackled, and it was Keith, the incident controller. 'BFO 3, can you come back to the control point asap?'

When I got there a few minutes later, he said, 'Greg, they're getting lots of 000 calls to Bantry Bay, and Comms seem to think it's probably just smoke from this fire. That doesn't sound right to me though, it's a fair distance from us. You know the area, so can you go and have a look?'

'Of course. No worries.'

I drove a few hundred metres, pulled over a rise and looked to the east. Unable to immediately comprehend what I was looking at, I grabbed my binoculars off the passenger seat. I was still in suburbia, but a bit over a kilometre ahead of me in dense bushland I could see a line of flames about 300 metres long roaring above the treetops ahead of the strong westerly wind and heading up a steep slope towards Wakehurst Parkway, Frenchs Forest.

How the bloody hell had it grown so big, with nobody yet in attendance? I wondered how many 000 calls had been received, callers reassured that they were looking at the Killarney Heights fire and told not to worry. An honest mistake, but it had put us behind the eight ball.

The fire was belching out rolling clouds of black, brown, and white smoke that looked like it was boiling into the sky and being blown by the wind towards Allambie Heights. Observing the smoke column usually told you exactly where the fire was headed. A flank of the fire was about to impact on properties on Bantry Bay Road, Frenchs Forest, but flank fires are always less intense with smaller flames, so it wasn't my immediate concern.

I knew that the huge smoke column concealed a storm of burning leaves, bark and small twigs being blasted upwards in the hot, buoyant convection column. They would rise to a certain height then fall to the ground slowly, starting new fires wherever they fell. The higher the temperature and the lower the relative humidity, the more likely the burning brands would stay alight in the convection column. The hot, dry, windy conditions were perfect for spot fires, and therefore a nightmare for firefighters. The intense fire was drawing in large amounts of air, reinforcing and strengthening the already gusty winds.

The main body of the fire was deep in bushland and could not be accessed safely given the flame height, spot fires, and rapid rate of fire spread. Getting off the road to try to confront such a fire would be suicidal – it was totally out of control. I later obtained footage from an ABC News helicopter that showed them guiding a bushfire brigade four-wheel drive to safety along fire trails behind Allambie Road. If the driver, somebody I'd known for years, Keith Bennett, had taken a wrong turn, he wouldn't have made it to safety. Even with his decades of experience he had been caught out by the sheer number of spot fires that seemed to come from nowhere as he tried to drive to a vantage point about one hundred metres behind homes. It was a close call, but thankfully Keith lived to tell the story.

The only possible strategy at a fire like this is 'defensive' – property protection, with fire trucks working from roadways where they can take shelter when the main firestorm arrives.

There were no fire trucks on the scene yet, and this fire obviously had huge potential. It would easily jump the two lanes of Wakehurst

Parkway, then enter Manly Dam Reserve, which was surrounded by dozens of homes and other buildings. The bush was very thick and the lead-up to summer had been dry. The bush was like a timebomb waiting to go off, and this was exactly the sort of explosion we'd hoped to avoid.

I knew that once the fire crossed Wakehurst Parkway it would burn downhill for a few hundred metres, slowing slightly, but given the strong winds, not much. It would then race up the slopes towards Allambie Heights, a bit over a kilometre away, gathering speed and intensity as it went.

Fires always burn much faster uphill: heat rises and the flames preheat the already dry scrub above, making the fire faster and more intense. It's almost like watching somebody push down an accelerator pedal; the increase in speed and intensity is visible, immediate and often frightening, even to an old hand. It also means that more embers are produced, even more spot fires start to break out, and fires can then leapfrog ahead of themselves as all of the small fires join up and produce even more embers. A cascading effect.

Firefighters work through a logical progression of strategies and tactics. Wherever possible they attack a fire quickly and directly with hoses in an attempt to extinguish it immediately and to limit the area burned. If this isn't possible, they then try to contain the fire, establishing a perimeter that can be controlled to stop further fire spread. Lastly, if the other strategies have failed or obviously won't succeed, it comes down to focusing solely on saving lives and property. There was no stopping this particular fire – it was out of control. All we could do was try to save lives and property. At least when it reached Allambie Road there would be no more continuous fuel, unless it spotted about half a kilometre into Allenby Park to the east, surrounded on all sides by homes.

I shuddered. I knew this area very well. There had been lots of fires here over the years, and a couple of times I'd been a bit scorched and had some narrow escapes. I remembered one fire where a mini fire-tornado happened up towards Seaforth as the fire came up a steep gully, another funnel. Small logs were sucked up into the twisting convection column and rained down onto the road, damaging fire trucks and forcing some of us to run for safety. Thankfully nobody was injured but it was a very scary 15 minutes.

At Allambie Heights, where the fire was headed, there were lots of homes backing onto the bush; a major facility – the Sunnyfield Association, which provided employment for people with cerebral palsy – thankfully closed because it was Saturday; and a couple of aged care homes, all of them right in the path of the flames but thankfully with wide fire breaks behind them. I couldn't see a single fire truck, and with so much to protect, and so little time before impact, my heart was pounding.

After absorbing all of this in less than a minute, I grabbed hold of my radio microphone. 'BFO 3. Red. Red. Red!' I'd had trouble getting through to the State Communication Centre earlier because there were fires all over Sydney and it seemed like everyone needed help at the same time. Radio messages are colour-coded, and Red is the highest priority. I could imagine the organised chaos as communications officers redistributed the dwindling firefighting resources across the region and looked at their situation maps and dispatch screens. By now they were probably seeking permission to recall off-duty firefighters and crew reserve fire engines that we kept for days just like this.

I asked for assistance, requesting at least ten fire engines as well as bushfire brigade support at Bantry Bay Road and Allambie Road where the fire would hit hardest. Communications told me that they would do the best they could, but resources were stretched thin across the entire Sydney Basin, with many fires burning in the hot, dry, windy conditions. That sick feeling again, my pulse racing, and my stomach squirming . . . How long would it take to get help to all of the exposed homes and facilities?

Even though I had only been promoted to officer rank recently, here I was in initial charge of a very large, very dangerous, fast-moving fire. I knew my stuff, but it's impossible to put out a fire and protect people when you have no fire trucks or firefighters. And right now, I knew that I had nothing. One of the basic adages of command training took hold – stay calm, think it through, and just concentrate on what can be done; not on what can't, because that's pointless.

I switched on my flashing lights and sirens and pushed through the traffic as fast as I could safely go, passing a couple of fire brigade pumpers and bushfire brigade tankers responding in the same direction as me. *Thank God*, I thought to myself.

I realised that District Officer King must have overheard my radio message and immediately redirected some trucks from the fire at Killarney Heights to the greater threat at Allambie Heights. Other crews would have been alerted and sent from throughout Sydney, but they would take longer to arrive. The bushfire brigade control centre at Terrey Hills had requested urgent out-of-area support and bushfire brigades were on their way from Sutherland and Baulkham Hills, at least an hour away.

The Killarney Heights fire wasn't over by a long shot, but the head of the fire had burned up to people's back fences and then run out of fuel. The flanks were still burning and I estimated the fire would take a day or two to round up. I knew that Keith would be releasing whatever resources he could spare as the threat eased at the first fire.

It took me about 10 minutes to get to Allambie Road through the thick, chaotic traffic. Police had already set up roadblocks and there was a lot of confusion. As I drove down the hill I saw pockets of black smoke and flames reaching into the sky close to homes. Spot fires already!

I had hoped that we might get about 30 minutes to set up and prepare before actually fighting the blaze, but realised that as the fire reached the bottom of the gully, it had intensified and accelerated even more than I'd anticipated. At one vantage point I could see at least ten new fires, small at first, but each of them rapidly gaining in intensity and starting to run towards homes.

I grabbed the radio again. 'BFO 3. Red. Red. Red!'

'Come in BFO 3 – pass your Red message.'

'On scene at Allambie Road, Allambie Heights. Numerous spot fires starting to impact houses, and two nursing homes about to be impacted. The main fire front is about 15 minutes away and is now on a 500-metre front. Will require at least fifteen additional appliances on this side of the fire and a senior officer to take command.'

My message was acknowledged and I heard sirens, then saw a few fire trucks pulling up with crews racing into action. Phew!

Some police cars and ambulances had pulled up at the nursing homes. Dazed, confused residents with blankets over their heads were being hurried outdoors, bundled into ambulances and whisked away, coughing in the thick smoke. Huge flames towered over the roof of one facility as a spot fire roared into the backyard – but the wide fire break stopped flames reaching the building.

A couple of buses arrived, evidently arranged by police to help with the evacuations. A crowd had gathered to watch the spectacle and merely managed to make nuisances of themselves. More people to worry about when the main fire front arrived, but at least they were on the other side of the road. Some of them stood in the middle of the road to get a better view or to take photos – causing emergency vehicles to swerve and jam on their brakes in the thick smoke. Bloody idiots!

It never ceases to amaze me how clueless some people can be. But I also know that the vast majority are sensible and helpful, and as an emergency responder you learn to deal with occasional bouts of incredulity and frustration with a simple shake of the head, since there's no point in getting angry. They don't mean any harm, and our job is to protect everyone – the clueless included.

I drove from scene to scene and lost track of time. There were now dozens of fire crews working hard to save homes, dragging hoses through backyards and knocking down flame fronts with high pressure hoses. I directed crews into driveways where homes were at risk and helped them run out hoses a couple of times.

The Sydney Fire District Inspector, Harry DeAudney, had arrived to take overall command and was in constant radio contact with me, building a picture of the situation and deploying resources as they arrived. Harry was the most senior fire brigade officer on shift in the state and was an old hand at bushfires. He was also probably one of the most popular senior officers in the brigade because he was friendly, treated everyone with respect, and was very competent. He gave me a roving commission and asked that I stay out in the field, not at the control point he had established.

After a lull in the wind, suddenly I noticed the smoke lifting and changing direction. A moderate south-easterly started to blow, seemingly from nowhere. A southerly wind shift was forecast for later, but the south-easterly shift was unexpected. Theoretically, the lowering temperature and higher humidity as the wind brought maritime air over the fire would lessen intensity – eventually. But, as firefighters know, it takes time for the increased moisture in the atmosphere to be absorbed by dead and living vegetation. This means that fires often continue to behave as if it is still 38°C with low humidity for an hour or more after a wind shift.

It is always a problem when a so-called southerly buster, a gale-force southerly wind change, arrives after a day of high temperatures – something that happens regularly on the Australian east coast. It is also why fire commanders always pay close attention to the northern flanks of a fire, because when a strong southerly arrives, the northern flank is transformed into the intense head of the fire. In the blink of an eye, a relatively benign fire can climb into the treetops and double its rate of spread. If you're caught in its path without somewhere to shelter and take refuge, you can die. Wind changes have killed many firefighters and other people, and destroyed many homes over the years.

The sudden wind shift turned the long north-western flank of the fire into the new, raging head. The bulk of our firefighting resources were on the eastern edge of the fire and would take some time to relocate, because the new fire perimeter was now several kilometres. Fires have the advantage of burning in a straight line; fire trucks have to follow winding roads. Fires often win the race.

I swung the LandCruiser around and headed back to Frenchs Forest. The fire was impacting properties in several streets and there were only a few fire trucks there so far. I radioed in the new information and asked that trucks be urgently reassigned to the new threat.

The smoke was incredibly thick – eyes burning, throat raw. I pulled into Parni Place and saw fire burning through people's gardens. The parents of my childhood friend, Phil Bush, with whom I would fight fires as a volunteer years later, lived on this street, but I could see that Bob and Bett were okay. A house further down the road backing into the bush looked like it may have caught alight, but because of the thick smoke it was impossible to tell.

I drove there, jumped out and ran down the driveway through wind-driven smoke and embers and noticed that a bushfire brigade four-wheel drive, a command unit, had just pulled up. Ted Carroll! An old friend and former captain of Terrey Hills Bushfire Brigade, now a volunteer group officer – a member of the command team for the Warringah Shire Bushfire Brigades.

Ted and I had fought lots of fires together over the years, including one in the mid-1970s in this same area. In my early years in the fire brigade, up until 1982, I remained a volunteer bushfire fighter on my days off. Ted was also a shift-worker, an air traffic controller at Sydney Airport, so we would often team up and respond to fires during the week.

We both did a 360-degree size-up around the house and found it was reasonably safe, although some external cladding was melting and sagging down due to the intense radiant heat as fire burned all around it. An elderly lady was gamely facing off against the flames with a garden hose, so I led her through the smoke to the front door, asking her to stay inside and take refuge. Ted grabbed the garden hose and kept the back deck wet as flames encroached.

Sirens. At last! The cavalry arriving in the form of multiple fire trucks. Ted and I agreed on a course of action, then went off to deploy our respective resources, him directing the orange trucks while I directed the red ones. There was no direct radio contact, but we made it work.

Then *another* wind shift. This time the expected southerly buster. The fire we were fighting started to die down and run out of fuel as it hit suburbia. Then I overheard a radio message: 'Pumper 51, respond to reports of bush alight, between Kens Road and Sorlie Road, Frenchs Forest.' I later found out that the fire had spotted about 1.5 kilometres to the north, across six lanes of Warringah Road, over the top of Forestway Shopping Centre and suburban homes, starting a new fire that would burn into Davidson Park alongside Middle Harbour. By this time, it was somebody else's problem though. I'd had more than enough for one very long and exhausting day, and Inspector DeAudney stood me down. It was a bit frustrating realising that fires were still burning, but it was vital to recognise fatigue before it became a problem.

As night fell I headed home, but was redirected to a fire burning in Ku-ring-gai Chase National Park between Bobbin Head and Mount Colah. I visited the control point briefly and made sure they had everything they needed. The bushfire brigade was in charge of this fire with the fire brigade assisting in a minor capacity. The fire was well within the National Park and no property was in danger.

Eventually I made it home and the kids peppered me with questions while Erris got me something to eat. I thought back to my own childhood and smiled to myself – just like when I was a kid and Dad would walk in red-eyed, blackened and smelling of smoke. What goes around, comes around.

It had been a big day, but as I'd predicted, the weather moderated and all of the fires were contained the next day. There was nothing particularly unusual about the fires, and as Dad had said to me earlier in the year, we were due for some bad fire weather and big fires because history showed that they are cyclical. It had been eleven years since the last big local fires, so the old patterns observed for many years had been reaffirmed again.

Up until the mid-1990s, fire seasons along the east coast of New South Wales had always been fairly predictable. In the hierarchy of bushfire threat, New South Wales was not considered as bad as Victoria, which historically had experienced far more loss of life and property from bushfires. Losses of homes in the worst years in

New South Wales were between 100 and 200, whereas major fires in Victoria in the past had destroyed ten times that. This was largely due to latitude, vegetation and topography, with Victoria being similar in many ways to fire-prone California.

In 1990, I had yet to hear of climate change, had no inkling that weather patterns were about to start changing significantly, and that the ability to predict how fire seasons would pan out would become far more difficult. The past would no longer be a good guide to what to expect in the future.

In 1991, I snared what I then considered to be my dream job, Sydney Region Hazard Reduction Officer / Assistant Bushfire Officer, but it didn't last. After acting in the position for a few months waiting for my appointment to be published, we received a call one day saying that the chief fire officer, Bill Reay, was coming to visit the Bushfire Section and wanted to address us all. Nothing like this had ever happened before and the team and I were apprehensive.

The chief didn't bring good tidings, and explained how, owing to budgetary issues as well as the growth of the new Department of Bushfire Services that coordinated bushfire brigades, it had been decided to cut out the three Fire Brigade hazard reduction teams that worked in Sydney, the Newcastle / Central Coast area, and the Blue Mountains.

The Bushfire Section would be reduced to just one district officer, two station officers, and two senior firefighters who would play more of a coordination role than hands-on hazard reduction burning. I was asked if I wanted to stay but knew that doing so would displace one of two older officers who loved the work just as much as I did. I had more time left in my career and more opportunities to get back into bushfire work so I decided I would go back on shift working at fire stations for the time being.

This was a major turning point in my career because I soon realised that I would have to study for promotion to district officer rank if I wanted to get back to the Bushfire Section – something that I probably wouldn't have done if I'd stayed in the Hazard Reduction Officer role.

Becoming a district officer meant that I would no longer be on shift at a fire station in charge of a crew of firefighters, one of the big attractions of staying at station officer rank. The exam was a big deal and had a very low pass rate. I studied for a solid six months and took five weeks' long service leave leading up to the series of written and oral exams. Thankfully I passed with flying colours, coming first, and promotion came far more quickly than I had anticipated.

Promotion opened up a lot of interesting doors, providing greater insight into how the organisation worked and how it could be improved. In any hierarchical structure the lower down you are, the harder it is to bring about change. I had lots of ideas and wanted to get things done, so after being promoted to district officer in 1992 (a rank that was later renamed inspector), I continued to study and learn. With the benefit of hindsight, I'm glad I did. I hadn't realised it yet, but the way we'd been neglecting and abusing the environment through ignorance, or greed (or frequently, both) was beginning to pay a terrible dividend. Fire seasons were about to get exponentially worse and bushfire prevention and response would have to change in ways we never imagined. Soon there would be no such thing as a normal summer. We would have to try to adapt.

Three decades on, I can look back at the bushfire season of 1990 – tough as it was in parts – with nostalgia for the good old days. It was one of the last years when fires were something we could predict, and something we could manage within our capabilities. I did not notice that the world was changing around me until the bushfire season of 1994, which made me realise that I hadn't even started to understand the scope of the threat we were facing. I began to question what I understood about fires and fire weather, and to look for answers.

3

THE PENNY DROPS

I CAN RECALL EXACTLY when the climate change penny started to drop for me. I'd been fighting fires for more than two decades as a volunteer and then a career firefighter. I had a good grasp of the weather, the land, the bush. Like all of my colleagues who'd cut their teeth fighting bushfires, experience had taught me how to anticipate what Mother Nature would throw at us each year, and there were lots of indicators to warn us when a bad summer was coming. This year, though, I was caught unawares. The world was changing around us, and somehow, despite small signs all around me, I didn't see this particularly bad fire season coming.

It was December 1993. For the first time in many years my annual leave fell over the Christmas period, so Erris and I had arranged a camping holiday at Gloucester Tops, north of Newcastle, close to the beautiful Barrington Tops wilderness. It was a place we'd visited as a family several times over the years, and we were excited to be having a break.

I was always nervous taking holidays in spring and summer, the bushfire season, because, like every firefighter, I felt a duty to be available when most needed. At the time, I was a district officer and had been filling in various roles when other officers were on leave, including back in my old haunt, the Bushfire Section, where I had spent a lot of time since 1984, most recently as acting officer in charge. When I wasn't doing shift work, I was on 24-hour call, and Christmas was, of course, often a time of serious fire weather. Erris, Phil and Kate had grown accustomed to me being called away at short notice and missing family gatherings, just as I had as a kid when Dad was often called away to fires.

This year was a bit different, though. We would be virtually uncontactable out in the bush, and it was very unlikely that my annual leave period would be disturbed, except in a dire emergency.

The other factor that made me less anxious was the prevailing wisdom among firefighters that this fire season was unlikely to be a significant one due to good rains in November. Each year I was alert to see what the weather would do in October and November, often a period of unstable weather in eastern New South Wales. Storms and rain in the lead-up to December often meant that it would stay quite humid until late January, meaning that dead and living plants absorbed more moisture and were therefore less prone to ignition. This would usually get us over the hump of the likelihood of serious fire weather, as it was rare for areas close to Sydney to experience major fires after January.

In the second half of 1993 I had attended several bushfires and noted early in the season that they were spreading quite quickly and were often generating spot fires at times and in locations that I wouldn't normally expect. Nothing too drastic, but just not quite right. There had been a bit of discussion between agencies about the fire behaviour, and it was agreed that it reflected that bushland was quite dry after a long period without rain. When it rained in November, though, the discussions moved on and most of us relaxed. We also discussed how all serious fire seasons in the past had been preceded by drought, and we weren't officially in drought at that time.

Up until the mid-1990s, fire seasons along the east coast of New South Wales had always been fairly predictable. If winter rains were scarce and spring was a bit warmer than usual, it was a good bet that there would be some bushfire activity through spring and summer, usually from October to February. The fire weather would start each year in Queensland, sometimes as early as mid-August, then shift progressively south as summer approached, meaning that there was usually plenty of warning that things might get interesting. As fire risk increased in one area, it would reduce in another, enabling us to shift around firefighting resources.

Weather was always the deciding factor. Rainfall, temperature, humidity, wind strength and direction all dictated levels of fire danger and the difficulty of getting fires under control. There are bushfires every year, but most years they are easily controlled, with just a few days where significant fire weather presents problems for firefighters. During the occasional hotter, drier years, particularly if accompanied by a drought and an El Niño event (see Appendix), there would inevitably be individual days when conditions became extreme: days when fires were hard to control and properties could be lost.

Throughout my years of fighting fires, and my dad's before me, the weather had behaved according to a certain set of rules. If it was going to give you trouble, it at least gave you fair warning. On the New South Wales coast around Sydney and west to the Blue Mountains, weather patterns leading to bad fire seasons happened about once a decade.

Dad had generally been able to pick the bad fire years as they approached, pointing out plants that were flowering early, or the early arrival of certain insects. He'd watch the weather patterns and how early season fires were behaving. But in late 1993 I remember him shaking his head and saying; 'I don't know. I can't pick this one mate. It could go either way.'

After spending Christmas Day with my parents and Boxing Day with Erris's, we set off from our home at Berowra, towing our camper trailer, on 27 December. It was an overcast day. I vividly recall, looking east over Ku-ring-gai Chase National Park, seeing a small bushfire.

Nothing to worry about, but it made me reflect that the effects of the November rain must be starting to wear off. I later found out that it was near Illawong Bay, my old territory not far from Terrey Hills. A couple of weeks later, the same area would be blackened by a fast-moving fire that would only stop when it ran out of fuel as it scorched its way to the ocean.

We were excited to be camping on the banks of the Gloucester River, a great base from which we could explore different parts of the huge wilderness area and mountains, or just spend time swimming. Phil and Kate loved the place, and Erris said that if she had the choice of a five-star resort or a tent by the river, the tent would always win hands-down. Another advantage of our chosen spot was that there was a small shop and an amenity block with hot showers nearby – luxury!

We pitched our tent among the trees within about 5 metres of the gurgling river as it tumbled over the rocks. Phil and Kate had a great time shooting the rapids in a small blow-up boat that Santa had given them, and swinging into a big waterhole using a rope that somebody had tied to an overhanging tree. At night we'd sit around the campfire and I'd do my best to make up ghost stories while we toasted marshmallows and gazed into the flames.

It was one of the best breaks we'd had in a long time, and the days seemed to last forever as we explored the area around the campsite, tried some fishing, and went up to Barrington Tops for longer walks. The area was stunning in its beauty – untouched wilderness stretching as far as the eye could see, teeming with wildlife. It was calming but also troubling, as it underlined how elsewhere we humans had modified the environment, often with little regard to the delicate balances of nature. We had no idea what was happening in the outside world, and didn't really care.

On 4 January 1994, we decided to drive to the top of the mountain range for a bit of a hike. As we climbed, a strong westerly wind began to stir, something that always made me a bit jumpy. The worst winds for fires are those that blow from the hot, dry centre of Australia, bringing high temperatures and low humidity (air moisture). From a vantage point high on the mountain, I could see two big smoke columns in the distance – probably more than 20 kilometres away and no threat to us, but I noted with concern that they were large, active fires, obviously out of control and probably inaccessible to firefighters. They would definitely get bigger.

I started to wonder what was happening outside our holiday bubble. Prior to leaving, I knew that there were large fires burning in the north of the state, particularly around Grafton. Not at all unusual at that time of year, but it looked like the fire weather was starting to extend further south; again, fairly normal, albeit later than usual.

We hiked down from the ridge top to a creek with a big waterfall and sat down to have a bite to eat and a drink. As I caught my breath, I suddenly realised I could smell bushfire smoke. I couldn't see a fire, but a small amount of wispy, white smoke was drifting over us from somewhere upwind. Best case scenario, it was smoke from a distant fire blown by the strong wind. Worst case, there was a small fire somewhere close to us and we were downwind – not a good place to be. Small fires can quickly become big fires when it's windy and dry.

I was worried for my family. The wind was strong enough that we could get caught if there was a fire close by. It was very rugged terrain and the bush was thick, dry and remote. We were probably 20 kilometres from the nearest house, and we had no way of calling for help. The only possible escape if a fire took off would be to clamber over rocks and try to jump into the creek, which would be difficult and dangerous given the steep, rocky terrain.

I convinced Erris that we needed to leave. Phil and Kate were disappointed as they had just spotted a big goanna climbing a tree and a couple of rock wallabies scampering up the rocks. But like me when I was a child, they knew that the smell of smoke wasn't a good thing when out in the bush.

Our campsite was at the base of the eastern side of the mountain range, and I realised as we drove back that we'd been protected from the westerly winds since arriving. At our campsite we could feel only a light breeze, but looking at clouds skating by high above, I realised that the wind up in the mountains was still strong. By now alarm bells were starting to ring inside my head because I realised that there must have been a significant shift in weather patterns. I went up to the shop and rang fire brigade HQ from the phone box.

What they told me stopped me in my tracks: a stubborn weather system had set in, bringing hot, dry weather, and many fires were burning. There had been some lightning strikes in mountainous areas sparking remote fires that were spreading under the influence of strong winds, with no hope of containment until the weather improved. They asked for a contact number as there had already been a serious drain on senior officer numbers required to resource incident management teams, and I might be needed.

I hung up, and sheepishly told Erris that I'd left the shop's phone number with work in case I was needed. She didn't say much verbally, but the narrowed eyes said it all! That night we tried to catch a news broadcast on the car radio, but we were too far out of range.

The next day was oppressively hot and we spent most of the time in the river. On the morning of 6 January, a bloke in a four-wheel drive pulled up near our tent and walked down to us.

'You need to ring the fire brigade in Sydney as soon as you can, mate,' he said. 'They asked me to come and find you.'

I drove up to the shop and called in. A chief superintendent, Doug Messenger, answered:

'Can you come back to work tomorrow? We need you to cover for D.O. North,' he said. D.O. North was the district officer in charge of fire stations in Sydney's northern suburbs.

We cleared up the campsite in record time and started on the road back to Sydney. I marvelled at my luck in marrying Erris. She just accepted that we had to leave, as did the kids. We found out years later that had we stayed, we would have been caught by a fire that swept down from the mountains into the campsite over the next couple of days, and we wouldn't have been able to get home because the Pacific Highway was closed on 7 January for several days as fires crossed the freeway north of Sydney. I speculated that the smoke I had detected the day before may have been a dead tree struck by lightning during a dry storm a couple of nights before.

As we headed south towards Sydney on The Bucketts Way, we were no longer shielded from the wind by the mountain range.

A hot, dry, westerly wind blew strongly, dislodging dead branches from trees, as leaves, twigs and dust buffeted the car. The temperature was in the high 30s.

To the south-west the sky was blotted out by a huge orange and brown smoke column in the distance – the Howes Valley fire, which would come perilously close to jumping the Hawkesbury River and burning through northern Sydney suburbs. The urgency of the situation struck me as we turned on the car radio and listened to news reports about the many bushfires burning across New South Wales and the forecast for serious fire weather over coming days.

On reaching home a couple of hours later, I confirmed by phone that I would be at work at 7:30 the next morning. They told me that a major fire had broken out in Lane Cove National Park and that I would probably be detailed as fireground commander if it was not contained overnight. I could see the thick smoke in the distance from our front yard.

As we unpacked, I thought about what we needed to do in case fires came near our place – we weren't in the bush, but we were close. Thick bush started 300 metres to the west, and there was national park all the way to the coast across the highway to our east. Well within range for wind-driven sparks from a bushfire, or ember attack, that could set fire to dry grass, bush and leaf litter. We had big eucalypt trees in our yard that had dropped a lot of bark, twigs and leaves. The grass was brown and had not been mown while we were away. I looked at it from a firey's perspective and saw a bonfire waiting to happen. I did a bit of cleaning up around the yard, mowed the lawn, put hoses out and blocked the downpipes on the roof gutters so that I could fill them with water.

Friday 7 January 1994 arrived. I hadn't slept well and headed to work early. The off-going district officer told me that they had contained the Lane Cove bushfire overnight but were still monitoring some burning trees on the perimeter, worried about another forecast day of Extreme Fire Danger. A Total Fire Ban was in force because it was going to be another hot, dry, windy day.

I jumped into the red fire brigade station wagon and drove to the incident control point at Lofberg Oval in Pymble. An old friend, Superintendent Bob Dobson, was the incident controller who briefed me on the incident action plan. Another close friend, Superintendent John Anderson, was a deputy incident controller at the Hornsby Fire Control Centre, where they were overseeing operations at this and another major fire to the north, working under Section 41F of the Bush Fires Act, which gave the controller sweeping powers to direct all agencies, including the police.

Given the forecast winds, the main focus would be on patrolling the fire perimeter and extinguishing every smouldering tree to eliminate ignition sources, then responding quickly to any new outbreaks. The big danger in hot, dry, windy conditions is embers – sparks that can carry for long distances, land in unburnt bush and start new fires. Dead trees, logs and stumps can burn inside for days or even weeks. 'Mopping up' is laborious, detailed, dirty work, and involves firefighters dragging hoses through burnt bush putting out every possible ignition source.

Throughout the day there were a number of flare-ups, but we got onto them all quickly. Later in the afternoon, the wind really picked up. There was a major outbreak near Macquarie University at Waterloo and Crimea Roads, and crews did an outstanding job of saving a number of townhouses.

But at about 4:30 pm, things really hit the fan. I received a radio message asking me to investigate reports of a fire at Khartoum Road, Macquarie Park. When I turned the red Holden station wagon around I saw a large column of brown and black smoke at a point where I knew the fire could run unimpeded to the east. Sparks from a burning dead tree in the valley we'd earlier been concerned about probably caused the new fire.

As I drove the short distance, sirens blaring, I radioed for all available units to go to that area. I saw a group of six lime-green fire trucks that had just arrived from Canberra on the side of the road. Realising they had no direct radio communications, I pulled over and told their strike team leader to follow me. Just prior to this, two full strike teams, a total of eight fire engines originally assigned to my fire, were urgently sent north to the Central Coast to protect properties being impacted by another fire that had taken a run up there. Bad timing. In total I had about twenty-five fire trucks available, but I knew that as the fire rapidly grew, there simply wouldn't be enough fire trucks or firefighters to protect the dozens of homes in the path of the fire.

As I pulled into Khartoum Road, rolling orange flames were roaring up the steep hillside from the valley below where the Lane Cove River meandered. Flames towered above a four-storey block of home units. A corner apartment on the ground floor of one block was already burning, with fire spreading to adjoining units above and to one side.

Three fire engines pulled in behind me, and I realised that an old friend, Jim Hamilton, was in charge of one of them. Years later Jim served beside me as deputy commissioner of Fire & Rescue NSW. The trucks were crewed by new recruits, still a few weeks from graduating as fully-fledged firefighters, and their instructors from the training college were the drivers and officers in charge. Today they would get a baptism of fire and an assignment that would normally require four or five fire engines with experienced crews. They did an outstanding job and stopped the fire spreading beyond the original unit. I made a mental note to write them up (for an award) when there was time, but all I could do at that stage was wave, then rush off to the next hot spot.

More fire trucks arrived. I put them to work but knew that we had to try to get ahead of the main fire front and deploy resources into narrow streets perched on steep slopes above the river valley. Lane Cove valley was surrounded by densely populated suburbs. Dozens of homes were about to be impacted by raging fires.

As I pulled onto Ryde Road, a busy six-lane thoroughfare, I was surprised to find no traffic, but quickly realised that police had blocked the road as flames soared high above the three northbound lanes. Small fires were breaking out on the eastern side and I saw a couple of larger columns of smoke where new fires were already starting to roar into the treetops, driven by the strong winds funnelling along the river valley.

As I headed north, I saw a lone figure cowering from the wind, smoke and embers on the centre traffic island. I pulled up and yelled at him to get in. I discovered that his name was Andrew Jacob and he was a freelance news photographer. He'd intended to hitch a ride to safety but became stranded by flames when the road closed. I put him to work reading my street directory.

As we turned into Lady Game Drive, we saw that the historic timber De Burghs Bridge structure was on fire below the modern concrete bridge. I had to make the call that it was too dangerous for firefighters to try to save. The old bridge was surrounded by thick bush, and I knew that lives would be in danger if anyone tried to pull hoses down there, with fire bearing down on them from the west and also from the valley below. It had to be left to burn; a hundred years of history, lost to the fire.

The fire had jumped Lady Game Drive at the corner of Ryde Road so there were a few anxious moments driving between the flames on both sides of the narrow road. Fires were rapidly intensifying, and huge flames were roaring into streets full of houses. Most of those streets had no fire engines in them yet.

As I sized up the rapidly spreading fire's potential and where it would go, I confirmed what I'd feared, that there weren't enough fire trucks and firefighters available to save everything. I made an urgent radio call for more help, but the expected answer came back: 'You're on your own.'

Many other properties were under simultaneous threat from fires elsewhere: the Blue Mountains, Central Coast, Sutherland, and soon, also in Warringah on the Northern Beaches. Every available fire truck from the fire brigade and bushfire brigades was out battling flames, and unsung National Parks rangers courageously did the best they could with their small striker fire units that held only about 600 litres of water.

I knew from the colour of the smoke ahead that houses were burning, and I prayed that the occupants had been able to escape. Albert Drive was the first street to lose homes to the flames, but a team of bushfire brigade tankers arrived just at the right moment together with a couple of reserve fire engines from the fire brigade, older trucks pressed back into service and crewed by firefighters urgently recalled to duty from their days off. The volunteers and career firefighters saved many homes just in that street.

On reaching the intersection of Lady Game Drive and Fiddens Wharf Road, I saw flames raging across the bitumen at several points up ahead, fire on both sides of the road. I stopped a car behind me and asked him to park across the road and not let anyone through until police arrived. I stressed to him that if anyone went through, they would probably die. The wide eyes told me that my stark message had been understood.

More desperate radio messages came through of homes now burning on Bradfield Road as towering flames raged up out of the valley and straight onto a row of houses. The focus now was only on saving lives; property protection became a secondary consideration. I can't explain how frustrating it was to have so much to protect, and so few resources to do it with. But I knew that every other fire commander at every other major fire was in exactly the same position. It was just a shit of a day.

I decided to see what could be done on the other side of the wall of flames. The wind dropped momentarily, and I saw an opportunity.

'I'm going to push through to the other side. Do you want to come with me and navigate?' I asked Andrew.

'Yes,' he replied. 'I'm coming.' He didn't think twice about it.

Then I accelerated through the smoke and flames down the narrow road surrounded on both sides by thick bush, squinting at the road to see if anyone was coming the other way.

Many people have died over the years after colliding with trees or other cars in the smoke. Hopefully, my flashing lights and siren helped ensure that other cars would know I was coming.

It was a calculated risk. I had to drive slowly enough in the dense smoke that I might have a chance to stop if I came across a fallen tree or a car coming the other way, but I also had to drive fairly quickly, as flames were licking my side of the car. I told Andrew to get as low as he could because the radiant heat through the right-hand side windows was literally blistering. I felt the skin on the right side of my face and right forearm burning. Later I would notice that I had sunburn and some blisters on that side of my face. A few days later the dead skin peeled off.

As we drove down the hill with fire on both sides I saw a line of cars up ahead stopped at the intersection of Grosvenor Road, the drivers obviously planning on heading up the hill but thankfully thinking twice. I ran over and said they all needed to turn around and get to safety. They didn't need much encouragement, as spot fires were breaking out all around us on the edge of the road.

One of these was right in front of me, already as big as a bus, flames leaping into the treetops while another fire started about 30 metres away, immediately burning up the steep hill towards a row of houses nestled in the bush above. I had no fire trucks available for this new threat and speculated on how many homes above would be destroyed.

I jumped into the car and said to Andrew, 'We need to be up there where the houses are – tell me where to turn off!' Using the street directory, he worked out that they were on Winchester Avenue. It took us a few minutes to get there and by the time we arrived the whole street seemed to be on fire.

Flames were leaping from the bush behind the houses, reaching high above the roofs; gardens were on fire, and at least four homes were already starting to burn. Soon, bright orange flames started to pulse out of upper storey windows of two of them, and black smoke poured out of the eaves and roof cavities of a couple of others. Power lines were down on the road at one spot, snaking around and spitting out deadly blue sparks. An explosion and the sound of breaking glass as the windows of a house blew out.

I picked up the radio microphone and sent another 'Red' priority radio message for assistance, saying that several houses were alight, but realising that additional help was unlikely. My concern was that every house in the street would eventually burn, as flames from burning houses encroached on others in a domino effect. In addition to this, a storm of embers was still being driven by near gale-force winds, and these could set fire to other homes further away from the bush interface. I parked on a bend in the road, looking down through a tunnel of fire, feeling powerless. The fire was too big, and we had too few resources. Again, I felt quite useless.

Three figures running up to the car caught my attention: two young police officers and a man, Barry, who looked desperate. He told me that his wife, Merja, was trapped in their home at the end of the cul de sac, and their back deck was already on fire. Flames had cut off any hope of escape and she was sheltering inside the house, as I later learned, sitting in a bathtub full of water. Unfortunately, that wouldn't help her once toxic smoke started to fill her house.

I looked at the street and realised that even wearing breathing apparatus, I probably wouldn't make it the 200 metres or so to the house. I was more worried about being electrocuted than burned with the live powerlines snaking around on the ground. I couldn't do it on foot. I would have to drive. I threw my breathing apparatus set on my back and tried to get behind the steering wheel of the car – of course that didn't work, because the air cylinder was too bulky. I was trying to reassess and figure out what to do when Andrew, who had been snapping pictures, ran up to the driver's side.

'Get in!' he yelled, and I bundled myself into the passenger seat as he got behind the steering wheel. Off we went.

We reached the house at end of the road and I jumped out. Flames were blasting across the driveway, blocking access to the house and the heat coming up the slope was overpowering. I waited for a lull in the flames, then ran through to the front door. Through the window beside the door I could see the back deck on fire and the house rapidly filling with smoke. A smoke alarm was screeching out its shrill warning. I tried the doorknob, found it was locked, then started banging on the door, readying my 'firefighter's master key' (my boot) to kick the door in if necessary.

Suddenly the door flew open to reveal a frightened but relieved looking woman – Merja. I shouted to her above the noise of the wind and flames to come with me.

'Not without my dog!' she insisted, so I went in, grabbed her ter-rified Keeshond, and we ran full pelt up the driveway through the smoke, dust and sparks. I looked back and saw the fire had entered the ground floor lounge room. We got out just in time.

As we reached the car I handed Merja her dog and pulled open the back door. Barry was sitting in the back seat. Confused, I did a double-take, wondering how he got there, trying to remember if he had come with us or not. No time to figure it out now.

'Get us out of here!' I yelled at Andrew, climbing into the front passenger seat. 'Quick!'

'But what about the two coppers?' he said.

'Huh? What coppers?'

'Those coppers.' Andrew pointed outside the car and explained that the young policemen I'd encountered up the street had made their way with Barry through people's yards unaffected by fire to reach us. Now, wearing only short-sleeve shirts, they were being blasted by smoke, heat and embers, and were lying down behind the car's wheels trying to escape the blistering gale. There was no room in the car.

'Onto the bonnet!' I yelled at them, climbing out to help them up. 'Now!'

They couldn't see in the smoke and needed help to get onto the car's bonnet. I told them to hang on to the windscreen wipers and not let go. They lay there as Andrew manoeuvred us through the bil-lowing brown smoke, past the arcing powerlines, through the sparks and flames. Finally, we were clear. An incredible sense of relief. We'd made it to safety.

By then a fire engine from Willoughby had arrived and the crew started to attack the northernmost house fire, trying to stop the home next in line from also catching alight. The embers raining down on us were so thick they'd set equipment on the fire engine's roof alight, and I ran to tell the pump operator. Next to arrive was a fire engine from Newtown Fire Station, then Headquarters' 'Flyer' (the first-response fire engine at Australia's busiest fire station which rarely leaves the inner city), then yet another from The Rocks. Reinforcements at last!

The Rocks crew went directly to Merja and Barry's home and, using just the water in the truck's 2000-litre water tank, knocked down the fire that by then was starting to engulf the lounge room and kitchen on the ground floor, and to spread upstairs. The hydrants soon ran dry. All up, we lost six houses in that street, and narrowly averted a tragedy. 'Bloody good firefighting by Station Officer John Roach and his crew,' I thought to myself.

Sometime later, bushfire brigade tankers from my old haunt, Warringah, arrived, a very welcome sight given the dry hydrants and that the fire engines were rapidly running out of water. George Sheppard, an old friend from the bushfire brigade, told me that they'd been in Albert Drive earlier and had saved a number of homes.

I left that scene and again tried to get ahead of the fire front, the hottest, fastest part that does the most damage. I began to deploy the extra resources that had started to arrive. By now just about every fire truck in Sydney was out among the flames, and off-duty firefighters had been recalled and placed on older reserve fire trucks.

As I drove up Delhi Road towards Chatswood, a powerline hanging down across the road loomed out of the smoke directly in front of us and there was no way we could stop in time. Andrew and I both squealed like children as we ran into it, blue sparks lighting up the smoke, then we were through. Afterwards, we laughed nervously and agreed to keep our less than stoic reaction to the live wire a secret.

The rest of the afternoon and night was a blur of action, with seventeen homes destroyed and about thirty others damaged, but dozens more were saved through the amazing, courageous efforts of firefighters. It was a similar tale of destruction in other parts of the state.

At one stage earlier in the afternoon I had radioed to see if we had access to any helicopters – one was available, so I arranged for it to land on a football oval to pick me up. I went for a quick flight to gather intelligence and develop an action plan. Looking north, I was confused. I could see the big fires up around Gosford about 70 km north of Sydney, but there was a huge pall of smoke much closer, obviously a very intense, fast-moving fire. Disoriented, I asked the pilot if he knew what it was and he said, 'Ku-ring-gai Chase National Park is burning. A fire started at Cottage Point and they're already evacuating Elanora Heights.'

I knew that my dad would be out among those flames and hoped that he and all the other firefighters were safe. Another thirty-four homes would be destroyed by that fire and over the next few days it would become the largest bushfire in recorded history to hit Sydney's Northern Beaches, despite a long history of major fires.

When I got back to the control point, Superintendent Dobson had left. He had been directed to take overall control of the Warringah fire which had also been declared a 41F Emergency under the Bush Fires Act. The replacement incident controller, Superintendent Ray Lee, asked me to stay and develop the over-night incident action plan because he was still trying to get his head around the situation.

I was worn out and had a throbbing headache from dehydration, but I drew up the plan, prepared a map, and briefed the incoming commanders. I put in a call to John Anderson at Hornsby Fire Control Centre to get approval for a backburn overnight, outlining the plan in detail as well as resource requirements. Ray Lee asked me to brief the local MP, Kerry Chikarovski, then I headed home.

All I wanted was a long, hot shower and a few hours of sleep before the next day, yet another day of Extreme Fire Danger. I tried to remember any time in the past that had seen so many consecutive days of serious fire weather.

When I handed over at about 11 pm I was told to take a spare fire brigade car home so that I could get through roadblocks and so that I could be called in early if needed. A couple of times I had to activate the flashing lights and police officers waved me through.

As I drove into Berowra just before midnight I saw dozens of cars and trucks parked all over the road, and hundreds of people milling around. Motorists and truck drivers trying to head north, stranded when fires blocked the highway. The sky glowed an angry orange to the north, and to the east another halo surmounted the now huge and menacing Warringah Fire, which could easily be on our doorstep if a strong southerly wind change eventually arrived. The motorists were safe, for now, as there were no fires to our west.

As I pulled into our street I saw all the lights in the house were on. A camper trailer was up on the footpath with a power lead running into our place. People – neighbours and strangers – were on our porch. I shouldn't have been surprised. My gorgeous wife was always finding ways to give people practical help. She had gone out to the families stranded near our house and offered food, drink, and use of our toilet and shower.

When I walked inside, blackened and smelling of smoke, I saw she'd been making sandwiches for stranded families.

I gave her a hug and said, 'All I want is a shower.'

'Well, you're number three,' she said. 'Sorry, but you'll have to wait your turn.'

I waited my turn, had my shower, then tried to get some sleep. For the next few days there were cars parked bumper to bumper along the highway, but most of the occupants moved on and found accommodation.

The fires went for days after that and I ended up fighting fires as far away as the Blue Mountains and was then placed in charge of a large NSW / ACT task force at Terrey Hills and St Ives.

More than 200 homes were destroyed, as well as many other buildings, with 800,000 hectares scorched. A number of people lost their lives. There would have been a lot more if not for the brave actions of firefighters from across New South Wales and from inter-state who mobilised and risked their lives to save those trapped in the path of fires.

I'll never forget when a Lane Cove homeowner led me through his garage to his back porch on 8 January. I looked down into the rear yard three metres below where a fire crew – from Beecroft fire station – were sitting on a scorched lawn, exhausted and black with ash. Behind them on the grass, which had been cooked dark brown by radiant heat from the towering flames that had just passed by, were three patches of green turf in the perfect shape of prone bodies. As the fire loomed over them, trapping them in the backyard, the crew had chosen (or so I thought) not to abandon the property but to stay and fight for it. They had lain on their stomachs, hosing each other to survive the intense heat. The image was almost comical, like something from a cartoon, but I was very much aware that the firefighters' lives had been in danger; that they stayed and put the safety of the community above their own.

I congratulated them and told them that I would be recommending all of them for a commendation. Marty, the crew leader, a bit of a character who sported an Elvis Presley hairdo, looked sheepishly at the others, then at me, shaking his head.

'Nah, boss,' he said. 'The bloke who owns the place closed the bloody garage door and we couldn't get out. I'd rather a beer than a commendation, but I'm on duty so that isn't going to happen. Where do you want us to go now?'

And off they went to the next job. The crew members all lived on the Central Coast and couldn't get home because of road and rail closures, so they slept at the fire station in between shifts over the next couple of days. They eventually received their well-deserved commendation, but I often think of that homeowner who either didn't know or didn't understand that he was putting the firefighters' lives at risk by shutting them out.

To date these had been the worst fires in New South Wales' history in terms of property loss. The 1994 fires were unprecedented, a word that is starting to lose some of its meaning as year after year bushfire seasons become longer and more destructive. Ever-worsening bushfire events have since become the norm, but 1994 was the year that really grabbed my attention. Importantly, not many people – even old firefighting hands like my father, who had been able to unerringly predict the worst bushfire seasons by observing nature and the weather – had not seen this one coming.

On reflection, and after lots of discussion with experts, it was clear that the defining factor was extreme and unexpected weather conditions. The weather was not behaving as it always had. In hindsight, when I realised that it was not just an outlier, this was when I started to understand that the weather, and climate, were definitely changing for the worse.

I believe that the 1994 fires were an early warning from Mother Nature. January 1994 was an early indicator of how climate change-driven extreme bushfire weather was going to continue to worsen as CO_2 and other greenhouse gas levels in the atmosphere continually increased.

4

SEARCHING FOR ANSWERS

THE 1994 BUSHFIRES WERE a worrying contrast to my understanding of a normal bad fire season and they started me on a voyage of discovery that would end up taking me around the world. Anecdotal evidence had reached me that firefighters in other bushfire-prone countries were becoming increasingly aware that weather patterns were changing. Old norms were no longer a given, levels of destruction by fires were increasing, and the dangerous, difficult jobs of firefighters were becoming ever harder and more perilous.

After my experiences in 1994 I wanted to learn more about what we might be able to change in order to better deal with increasingly large and destructive bushfires. I had yet to make the definitive climate change connection in my mind, even though I sensed that things were different, with weather patterns leading to serious bushfire weather subtly changing.

I applied for a Fellowship from the Winston Churchill Memorial Trust to fund an overseas study tour, because government agencies are chronically underfunded and rarely have spare cash for R&D, particularly for overseas travel. There was no chance whatsoever that I would get an opportunity to do an overseas study through 'official channels', which was just a reality of life. My application focused on the use of water-bombing aircraft and incident management systems to help control bushfires, and ultimately I was successful. My itinerary would take me to England, France, Spain, Canada and the USA for ten weeks from July 1995. It was a chance to speak to bushfire experts and study techniques they were deploying to deal with increasing bushfire problems in the hope that I would learn things that might assist us in Australia.

The Fellowship was not without its dramas. After being hosted by the London Fire Brigade for a few days, my visit to France coincided with protests from the Australian and New Zealand governments against continued nuclear testing by the French at Moruroa Atoll in the South Pacific.

On arriving in Marseilles late at night after a long train trip from Paris, I found a message waiting for me from my hosts saying that I should not come to the firefighting airbase in the morning as originally planned. The French Government, who saw me as a government official, had barred me from entering any government installation as part of its retaliation against the Australian Government's strident criticism of France's nuclear testing program.

I received several late-night phone calls from media outlets, and an Australian senator. Ultimately, after a few awkward days in which I was able to speak to key people but not enter the airbase, it was strongly suggested that I should head to Spain a few days early.

Discussions with Spanish firefighting authorities were very enlightening. Their aerial firefighting methodology – deployed in France and Canada as well – stemmed from a simple truism: that all big fires start small. The objective was to keep them that way, and they used the only purpose-built firefighting aircraft in the world, the CL-415, which is capable of refilling by scooping water from rivers and lakes. They were used very effectively in tandem with ground firefighting crews to isolate and control fires before they grew out of control. These agile firefighting strategies and tactics were subtly different to how aircraft were being used in the USA, and I could see how they would be better suited to Australian conditions. Despite their firefighting capabilities, the Spanish had also noted an increase in the size and intensity of fires, driven by worsening fire weather. Spain experienced major fires in 1993, as had many other southern European countries. Authorities told me that the fires had been far more widespread and intense than in the past.

I then went to the USA and visited the National Interagency Fire Center in Boise, Idaho, to learn about national coordination. The Americans had a different approach, in that they used medium and large re-purposed military aircraft, such as Lockheed C-130 Hercules and smaller S2 aircraft, to drop retardant substances ahead of the flames to create fire breaks, in tandem with water-bombing helicopters. The familiar red-coloured retardant was very effective in grasslands, brush, and conifer forests common across the western states, but not as effective in broadleaf forests because the leaves reduced the amount of retardant able to coat fuels on the ground.

Boise was also a base for 'smokejumpers': firefighters who are dropped into remote fires – such as those sparked by remote lightning strikes – by parachute. The smokejumpers would camp out and use hand tools and small portable pumps to try to keep fires small. Californian fire authorities were also using the CL-415 'Super Scooper' amphibious firefighting aircraft commonly used in Canada and Europe for fast-attack. The difference in aircraft strategies is that the Scoopers drop water and sometimes foam directly onto flames, immediately reducing fire intensity so that ground crews can directly attack fires, whereas the large retardant carrying aircraft generally drop ahead of fires to create fire breaks. The Scoopers had much better turnaround times between drops because they didn't have to land, refill, then take off again. All they needed was a suitable body of water to skim across. Even though smaller in carrying capacity, they were considered more productive than the larger air tankers, because they could quickly refill and return to fires, flying more missions and dropping more over time than the larger aircraft.

A study for the US Forest Service in 2012 recommended that the optimal aerial firefighting fleet should comprise a large number of amphibious water-scooping aircraft as the firefighting mainstay, backed up by a small number of large and very large air tankers for specific tasks and in specific vegetation types where they are most effective.

I spent more than a month in California with the Los Angeles City, Los Angeles County, Oakland, San Francisco and Sacramento fire departments. Most of that time was in Los Angeles, and I stayed at fire stations in Hollywood, South Central LA and the San Fernando Valley, responding to many different types of emergencies, including shootings, stabbings, household fires, factory fires, road accidents and bushfires. US firefighters were also medical first responders and many fire departments were hybrid fire / ambulance services.

The level of violence in LA at that time was stunning to me and I was issued a bullet-proof vest to wear when responding to medical calls involving firearms. One incident where two teenage brothers were gunned down in broad daylight while walking with their father in South Central LA will stay with me forever. I performed CPR on one of the young men, to no avail.

My chaperone during the LA visit, Captain James Featherstone, explained to me the depth of social problems in parts of LA and the USA, and how gun and gang violence had actually improved compared to the 1970s. I was appalled to learn that the Los Angeles City Fire Department had operated segregated fire stations up until about 1975, with all-black and all-white fire crews at different fire stations, with most people of officer rank being white. When I expressed shock, James cautioned me in my outrage, telling me that as an African American he had studied the plight of other black people around the world and Australia's treatment of First Nations people was nothing to be proud of. Of course, he was right.

James later became the head of emergency management for Los Angeles City, after which he served a term as interim Fire Chief. I consider him to be one of the most forward-thinking people in emergency management globally, and he shares concerns about climate change and all of its impacts. California is also on the front line of climate change, having to deal with escalating drought, heatwaves, desertification, wildfires, floods and landslides. We remain great friends and in March 2020 he arranged for me to address a large forum in LA on climate and fires. Sadly the conference was an early casualty of the escalating COVID-19 pandemic.

Early one morning during my 1995 study, around 2 am, I responded to a brush fire from an LA County Fire Department fire station in the San Fernando Valley. The First Alarm assignment (the initial dispatch at the time of call) was five fire engines with twenty firefighters, a bulk water tanker, a low-loader semi-trailer with a bulldozer, a busload of eleven firefighters from the nearest fire camp, and another crew from the camp transported by helicopter. A battalion chief (equivalent to an inspector) also responded, to take charge and manage operations. Strong Santa Ana winds, feared throughout Southern California, were expected the next morning, so it was imperative that all fires were contained as quickly as possible.

It was fascinating to watch the operation unfold. The fire was burning quite fiercely uphill, having started on the side of a mountain road. There was little wind at that time but the long grass, chaparral scrub and Scotch thistle brush interspersed with eucalypt trees was very dry.

The battalion chief immediately radioed for assistance, a Second Alarm, which was a doubling of assigned resources. He positioned two fire engines, one on each flank of the fire, and fire-fighters started to work uphill, stretching an ever-lengthening line of hose and extinguishing the flanks of the fire as they went. Each of these engines was backed up by another, which supplied water, and the fifth engine started to shuttle water from the nearest hydrant, about a kilometre away, until the bulk tankers arrived. Two buses arrived from the fire camp and firefighters with hand tools immediately started cutting trails about 2 metres wide behind each flank of the fire, following the hose line crews uphill.

I saw a helicopter arrive and hover at the crest of an adjacent hill that was not affected by fire, and more firefighters with Pulaski tools (a cross between an axe and a mattock) jumped about half a metre to the ground and headed for the brow of the hillside that was burning. The battalion chief explained that they would cut a trail across the top of the hill in an attempt to stop the fire there. This worried me, as above a fire is never a good place to be, but the chief assured me that they had their escape routes well planned.

Soon after this, the chief, acting on an updated weather forecast, assigned the bulldozer to cut a one-blade-wide break on the downwind side of where the hand tool crews were cutting their uphill trail. He explained that this was a fallback strategy and added safety factor given the possibility of strong winds later in the day.

The operation went with military precision and was incredible to watch. The fire was contained within two hours, before the dreaded winds started to blow from the desert to our east. I marvelled (not without envy) at the budgets of the two largest fire departments in Los Angeles compared to fire services in Australia.

Each had their own fleet of helicopters and there was a huge resource base of firefighters and fire engines.

Los Angeles County Fire Department had huge areas of wildland to look after and was well practised in dealing with brush fires. Each summer the Department hires seasonal firefighters, who live in barracks at facilities placed strategically throughout fire-prone mountain areas – the 'fire camps' can dispatch crews at a moment's notice to wherever they're required in the county. They are run on a military schedule, with strict daily routines starting with a morning run, group exercises, then breakfast in the mess hall followed by training drills.

While visiting Camp 2 near Pasadena one day, a siren suddenly sounded. The public address system announced that a crew had to immediately respond by helicopter to a reported brush fire, and another by bus. A crew of seven scrambled into a helicopter, which immediately took off, and another crew piled onto a red bus fitted with flashing lights and sirens, and off they went too. There was a small column of smoke visible in the distance.

Later that afternoon, the siren blared again. My battalion chief host, Bob, excused himself and hurried to the operations room. He came back soon after and asked, 'Want to go for a bit of a drive?'

Bob told me that there were two major brush fires burning to the east of San Diego, and two camp crews had been requested under the Californian Fire and Rescue Mutual Aid System, whose basic concept is that adjacent or neighbouring fire agencies will assist each other. We set out in convoy, Bob's red command SUV in front, and two busloads of firefighters behind us.

Bob and I got talking, and he confirmed that he had noticed that fires were changing in California. He mentioned the 1993 firestorm and told me it was getting hotter, fire seasons were lengthening, and droughts were becoming more frequent and far more severe. He was a veteran of more than thirty years on the front line, and, like me, he'd noticed changes, but as yet could not quite put his finger on exactly what was happening.

'Brush fires are one thing, but we have an introduced tree species that you really don't want to know about here,' he told me earnestly. 'They're called eucalyptus trees. Man, they suck! And they cause something we call spot fires.'

I broke the news to Bob that eucalypts were Australia's gift to the USA, and where I came from we had millions upon millions of acres of them. He said, 'Well, I guess we'll have to forgive you for that, but . . . sheesh.'

On approach to San Diego County, Bob changed onto the state-wide command radio network and received instructions. We headed into a mountainous area with scattered homes and thick brush. A plume of smoke loomed ahead of us.

What followed was amazing to watch. In an area the size of two football fields, a base camp was quickly established. Semitrailers hauling command units, catering units, shower and toilet blocks started to arrive. We had to go back out onto the road as they set up. On re-entering, I was issued with a firefighting uniform, tent, bed roll, sleeping bag, backpack and an ID card with a barcode.

Later, a number of black-and-white SUVs arrived with what I thought were sheriff's deputies aboard, but Bob told me they were officers from the California Department of Corrections. Soon after that, buses arrived filled with firefighters dressed in bright orange. Bob explained that they were prison inmate fire crews – trusted non-violent prisoners, some of whom were veterans of more than a decade of cutting fire line.

They were marched out to their respective areas – male and female crews kept well apart – and instructed to set up their tents while loops of razor wire were dispersed from the back of a ute, encircling their accommodation area. Armed guards watched their every move.

Within two hours, a small town had been established in the middle of nowhere. We would stay there for three days, until the fire was contained and the LA County Fire Department crews were released.

The crews only worked during the day, because there were too many hazards on the rocky mountainsides, including lots of rattlesnakes. Water bombing aircraft wheeled above us and strike teams, each comprising five fire engines and a command car, started to congregate out on the road. Hot shot crews from places like Alaska, Florida and Colorado arrived – well-drilled crews who, like the LA County camp crews, specialised in cutting fire line.

I was amazed when a caravan arrived and set up outside the camp entrance selling T-shirts, complete with the name of the fire. There was always a line of firefighters out there in the mornings and afternoons and Bob told me that this was a common thing, with the T-shirts being prized mementoes. Of course, I bought one too.

Next stop after Los Angeles were visits to the San Francisco and Oakland Fire Departments. On arrival in Oakland I was met by Captain (later Battalion Chief) Bill Wittmer. I didn't know it at the time, but Bill and his wife Cindy would become lifelong friends.

Bill arranged for me to go up in a police helicopter to look over the area where the 1991 Oakland Hills Fire had killed twenty-five people and destroyed more than 3,000 homes, describing to me how a small fire had been extinguished the day before the Oakland Hills firestorm and roared back into life with the arrival of strong winds.

It raced up the steep, wooded, heavily populated hills, outpacing fire engines that had to negotiate narrow winding streets clogged with evacuees coming the other way, further slowing their response. Bill told me how in street after hillside street every home and apartment block was on fire and there was nowhere for anybody to take shelter as heat blasted from below. People were killed on foot, in their cars and in their homes.

An intelligent and well-read man, Bill also expressed the view that Californian weather and wildfires were worsening, and pointed me towards early scientific papers on climate change. Over the following days he showed me every aspect of the Oakland Fire Department and its fire-prevention strategies.

One of the most innovative was to deploy an elite, mercenary specialist group – a herd of goats! People who had heavy fuel loads on their land could hire the goat man to drive his herd to the area in question, set up temporary fencing, and let them start munching. It made for very effective fuel reduction.

One of the key facts that my American counterparts reinforced was that changing weather patterns are not the only bushfire risk caused by climate change. As ecosystems that developed over millions of years start to break down as the environment degrades, so too will the delicate balances that maintained those environments.

A climate change impact unique to the west coast of the USA, not yet apparent during my 1995 visit, has been a large-scale dying-off of trees due to a parasitic insect that burrows into tree trunks – the western bark beetle. In the past, cold winters and snow would kill off the beetles each year and trees would recover. Warmer temperatures, drought-stressed trees, and less snow have had a devastating impact on forests, with millions of acres of trees dying because the beetles continue to feed and multiply in winter, killing the trees. Falling dead foliage and branches have drastically increased fire fuel levels and flammability, leading to faster moving, more intense fires.

I returned home to Australia with a deep respect for the bravery, professionalism and innovation of firefighters confronting escalating fire threats across the world. With that respect came a growing awareness that patterns I'd been noticing in Australia were not confined to our shores. We were looking at a global problem, and I could see no easy solutions.

5

THE QUICKENING: FROM BAD TO WORSE

AFTER THE 1994 BUSHFIRES there was an assumption by many within the fire services, me included, that there would not be another bad fire season in New South Wales, or in the Blue Mountains in particular, for another decade or so. This was the long-term expectation of the return period of weather patterns resulting in serious fire seasons, established over two centuries of observations.

Alarmingly, though, serious fires returned to New South Wales in 1997, when homes were destroyed in the southern suburbs of Sydney and two firefighters lost their lives near Lithgow on the western slopes of the Blue Mountains. I had a close call myself during those fires.

I was flying in a helicopter over southern Sydney when we were hit by a strong southerly wind change. Suddenly we were enveloped in smoke and the pilot lost visibility. We were flying VFR (Visual Flight Rules) and I realised that we were in a serious situation because the pilot could not rely on instruments to work out where we were. The fact that we were being thrown around like a cork in the ocean didn't worry me as much as the fact that we had no idea of our location and couldn't see the ground.

Bankstown Airport radioed to advise that we were encroaching on its airspace, meaning we had been blown several kilometres north. The pilot told me and the Rural Fire Service (RFS) officer with me that he was going to have to land, and that we 'might go in hard'. Splendid, I thought. We braced as he lost altitude and then suddenly we saw treetops next to us. He pulled up sharply to clear them, then skimmed above the trees until he saw a clearing and immediately landed – right in the middle of Heathcote Road – normally a busy thoroughfare, but thankfully closed because of the fire. We jumped out and ran in opposite directions to stop any oncoming fire trucks. After a while the wind died down and we took off again. If the road had been open I would have called a taxi instead – I'd had my fill of helicopter travel that day.

The 1997 fires were significant, not because of their intensity, but because the return of serious fire weather so soon after the 1994 fires was quite unexpected.

On Christmas Eve 2001, when I was Acting Commissioner, a fire started in the foothills of the Blue Mountains during hot, dry, windy conditions. At the State Operations Centre of the Rural Fire Service that night, RFS Commissioner Phil Koperberg, the head of the NSW National Parks and Wildlife Service, Brian Gilligan, and I realised that the next day would be testing.

The forecast for Christmas Day was for Extreme Fire Danger. Strong winds from the west would almost certainly push existing fires into residential areas while creating ideal conditions for new fires to start as the winds caused power transmission lines to clash, and branches and trees to fall across them causing arcing and sparks that could start new fires. Cigarettes discarded from cars could also cause new outbreaks in dry grass alongside roads and highways. While the immediate threat to life and property had reduced in the lower Blue Mountains as the fire moved into inaccessible terrain overnight, there was no prospect of containing that fire before the strong winds started early the following day.

I thought to myself that things could not get any worse. But as Murphy's Law sometimes dictates, they did. I woke at about 4:30 am on Christmas Day to the sound of thunder. I immediately thought, 'Thank God – rain!' I was wrong.

A thunderstorm that produced no rain moved over parts of the Blue Mountains and western Sydney, its lightning sparking several new fires. The worst would prove to be the Mount Hall fire that started in a remote part of the Blue Mountains National Park wilderness west of Warragamba Dam, Sydney's main water supply, and the township of Warragamba. Under near gale-force westerly winds, the Mount Hall fire forged its way to the east and easily jumped about a kilometre across the dam, destroying many properties at Warragamba and the neighbouring town of Silverdale.

The commissioner of the NSW Fire Brigades at the time, Ian MacDougall, was spending Christmas at his holiday house near Ulladulla on the south coast, hoping to have a couple of weeks off. His house came under serious threat from one of the many fast-moving bushfires in the state that day.

I called him, realising from the weather forecasts and the extent of the fires that we were probably in for a bushfire campaign – a long-running battle that could go for days or even weeks. We agreed that he should return to Sydney to take overall charge of fire brigade business as usual, enabling me to concentrate on the bushfires and coordinating strategies and resources in my normal role of Director of State Operations. The trip took him many hours as he dodged fallen trees and negotiated his way through police roadblocks.

While the Blue Mountains were burning, other fires broke out in the Sutherland / Illawarra area south of Sydney; impacted severely on the township of Helensburgh, at Brooms Head on the state's north coast; and at Wollondilly in Sydney's south-west.

On New Year's Day 2002, a new fire broke out in Lane Cove National Park close to the Sydney CBD where I had led firefighting operations back in 1994. I had just attended the Wollondilly Fire Control Centre south-west of Sydney with Commissioner Koperberg, and on the way back to Sydney diverted to the new fire, by now burning fiercely in the middle of suburbia, the river valley surrounded on both sides by hundreds of homes. I was interested to see whether the fire would take a similar path to the 1994 fire and I thought that I might be able to assist the Incident Controller with intelligence from that earlier fire, when many homes had been damaged and destroyed.

What I found most remarkable was that this time there was a decisive new factor in the firefight. Following the 1994 bushfires, an enterprising fire brigade station officer in the Bushfire Section, Terry Munsey, had developed the concept of Community Fire Units (CFU).

CFUs are street-based volunteer units composed of residents provided with portable pumps, hydrant gear and fire hoses, as well as distinctive blue overalls and protective equipment. Local fire station crews provide regular training in bushfire safety and use of the equipment. The volunteers are not taught to be firefighters, but to wet down the bush and prepare homes prior to fire fronts arriving, to take shelter when fires burn through, then to emerge afterwards to put out small fires and deal with embers that may otherwise destroy their homes.

The CFU program proved to be decisive at Lane Cove. Back in January 1994 I had run out of fire trucks as the fire spread quickly to the east. The lengthening flanks of the fire had to be resourced for hours after the fire went through because the many burning trees and logs generated embers that could easily start new fires in the bush or set homes alight. Eventually there came a time when there were no more fire crews available to keep pace with and knock down the intense fire fronts. Homes then started to burn.

This time, however, firefighters were able to hand over to the CFUs and keep ahead of the intense fire fronts. Despite it being an intense, fast-moving fire, no properties were lost.

The fires would ultimately destroy 109 homes and dozens of factories, shops, and community, farm and other buildings. The 2001–02 bushfire season would prove to be one of the longest in NSW, with fires burning for months. Happening just seven years after the 1994 fires, and four years after 1997, these fires were unexpected.

Still clinging doggedly to past experiences and paradigms of what to expect in terms of fire weather, nobody really expected that the summer of 2002 would see a return of major fires. To the contrary, we fully expected that it would be a benign season. However, the Millennium Drought was underway, and fuels remained critically dry.

On 7 December 2002 critical fire weather saw major fires start simultaneously across New South Wales: in north-west Sydney at Maroota; the Megalong Valley and Mount Riverview in the Blue Mountains; and Mangrove Mountain on the Central Coast. At the same time, a number of fast-moving grass fires broke out in western Sydney suburbs.

Yet again, the state was in bushfire crisis mode and a number of homes were destroyed as the Maroota fire headed east, driven by 40–60 kilometre-per-hour westerly winds. In its path were the suburbs of Dural, Glenorie and Arcadia. If it jumped Berowra Creek it would impact heavily on the suburbs of Berowra and Cowan.

I remember going to Berowra, where we had previously lived for ten years, heading to a vantage point on a steep wooded hillside with John Anderson, now the North Region commander and an assistant commissioner. In the dark we watched huge flames cresting hills on the western side of Berowra Creek, several kilometres away as the crow flies. The flames towered above the treetops and masses of embers blew on strong winds ahead of the fire, hopefully not reaching our side of the creek far below.

The creek, although about 200 metres wide, would prove to be no barrier to the mass of burning embers. As we saw some large embers floating down nearby, we realised it was inevitable that new fires would start below us. We asked a police officer to evacuate the lookout and keep people away, as by now dozens of people had arrived to look at the spectacle of the advancing fire and the huge orange glow in the early evening sky.

Not long after, we saw a separate orange glow somewhere on the foreshore way below us, an inaccessible area with no fire breaks between the creek and hundreds of homes nestled in the bush on the top of steep ridges above – a firefighter's nightmare.

John and I agreed at the time that we could expect heavy property losses. Incredibly, however, the efforts of hundreds of firefighters from the RFS, Fire Brigade and National Parks over the next few days saved every home as the fire impacted different areas at different times.

The following evening, I left the RFS State Operations Centre after a long day and went back to Berowra to observe operations and assess the impact. I could hear flames roaring out of a gully as cascades of embers and endless clouds of heavy smoke engulfed a row of houses. I ran into Superintendent Shane Fitzsimmons from the RFS. Shane had been working long hours at the State Operations Centre but had raced home to Berowra when he realised his own street was coming under threat. As flames advanced up the steep hillside below his home and others, he was there dragging hoses and battling the flames together with dozens of other firefighters. Years later he would become RFS Commissioner and lead the fight against the Black Summer fires.

Unfortunately, massive fires, started by lightning, burned through about 2 million hectares in alpine areas on the New South Wales / Victoria border in late 2002 and early 2003. The event the dawning realisation of mine that the frequency and length of major fire seasons was changing.

On 8 January 2003, lightning from dry thunderstorms in remote mountainous areas to the west of Canberra started at least four bushfires. Over subsequent days the fires grew, despite efforts at control.

On 18 January 2003, Extreme Fire Danger was forecast for Canberra. The familiar trifecta of high temperatures, strong winds and low humidity resulted in the now large fires burning into the suburbs of Canberra, simultaneously impacting homes and buildings on the urban / bushland interface over a front of about 12 kilometres. Fire services were totally overwhelmed, and a number of fire trucks caught fire in conditions accurately described by residents and firefighters as a 'firestorm'.

That day, Canberra gained the dubious honour of being the site of one of the world's first recorded fire tornados, with 487 homes destroyed and four people losing their lives. The extreme heat and convection from converging fires combined to generate a tornado that cut a 200-metre-wide swathe through a pine plantation. Phil Cheney, an experienced fire scientist, later reported to the Coronial Inquiry into the Canberra fires that wind strengths within the tornado were estimated to be in excess of 200 kilometres per hour. A sustained rate of fire spread of 20 kilometres per hour was the 'fastest documented rate of spread of a forest fire anywhere to my knowledge'.[2]

So, what does a fire tornado look like? Firefighters described the sky turning black in the late afternoon, so dark that streetlights came on automatically. Fully grown trees were snapped in half like toothpicks and strewn around for hundreds of metres, while roofs were ripped from houses that subsequently burned to the ground after a storm of burning embers set fire to furniture inside. One fire crew sought refuge below ground in a sewer as their fire engine blazed on the street above.

Authorities were shocked because while the Australian Capital Territory had experienced major bushfires before, it had not experienced significant property losses since 1952, and never on the scale experienced in 2003.

The alarming and incredibly dangerous atmospheric phenomenon of the Canberra fires overshadowed the 2003 Alpine Fires on the New South Wales / Victoria border that burned nearly 2 million hectares. In hindsight, the Canberra fires were an early warning of how a warming, drying landscape and gradual worsening of fire weather conditions provided a glimpse of a more dangerous and volatile future. It certainly gained my attention and that of other fire chiefs nationally and internationally.

Two major changes emerged following the Canberra fires, which captured the attention of the national government based there. Prime Minister John Howard agreed to a cost-sharing agreement between the states and territories to enable leasing of a larger range of firefighting aircraft, and perhaps the most far-reaching change, funding was provided for the establishment of the Bushfire Cooperative Research Centre (BFCRC) that operated until 2014. The BFCRC partnered with firefighting agencies to conduct crucial research that helped authorities to understand the changes that were occurring and assist in developing responses with positive improvements, such as better fire truck protection systems including sprinklers, better firefighting uniforms, insights into decision-making models in crises, and better predictive models which took into account worsening weather and fire conditions.

After the 2002 fires, many people again assumed that it would be years, hopefully about a decade, before we would again have to worry about significant fires and bad fire weather. The pattern of shorter periods between fire seasons and severe fire weather continued. Significant bushfires broke out again in New South Wales in September 2006, and homes were lost to fires in south-western Sydney. Bushfire emergencies were also declared in Newcastle / Lake Macquarie north of Sydney, the Hawkesbury, Baulkham Hills, and the Shoalhaven – on the New South Wales South Coast. This was the earliest declaration of a fire emergency, in the state's history, before the official start of the statutory bushfire danger period on 1 October. This reinforced the disturbing trend towards both an earlier onset of serious fire weather and the lengthening of fire seasons, something that would become well established in following years.

Other states were also confronted by heightened bushfire activity, with Victoria experiencing sixty-nine days of major fires in alpine areas that would burn more than 1 million hectares in the summer of 2006–07; homes were lost to flames in Tasmania, and major fires broke out in both South Australia and Western Australia. What this signalled to me and other experienced firefighters was that increasing bushfire frequency and severity could no longer be dismissed as one-offs. The numbers didn't lie. There was now a discernible trend, and the escalation was being driven by different weather patterns. The climate was changing. In my role at the time as a board member of AFAC, I was involved in a lot of discussions about the ramifications for fire and emergency services about the now obvious and increasing fire risks due to changes in climate.

On 13 November 2006, lightning sparked fires in the Blue Mountains National Park, eventually blanketing Sydney in choking smoke. Extreme fire weather drove a massive fire through the Grose Valley, leading to the mobilisation of dozens of fire engines from fire brigade stations throughout Sydney and the Blue Mountains, and dozens of RFS tankers from across the Sydney Region. At one stage the intense heat from the fire formed a pyrocumulonimbus cloud – a fire-generated storm – which resulted in extreme fire behaviour. This was significant, because the coupling of a fire with the upper atmosphere in this way, while not unheard of, was still a relatively rare and notable occurrence, but one that would become increasingly common as the atmosphere warmed and became more unstable.

Thankfully, the main fire did not impact on populated areas and stayed largely in remote, inaccessible gullies. This was not the first time that luck played a significant part in the ultimate outcome. But it was more than just luck. Remote area firefighters from the NSW National Parks and Wildlife Service played a critical role in bringing this fire under control. They worked in a race against time and weather to manually cut fire breaks in areas where the only possible escape was by helicopter extraction. They were the unheralded heroes of this massive, dangerous fire.

During the fire, Phil Koperberg and I met at the Blue Mountains Fire Control Centre and discussed strategies, considering a range of scenarios including possible wind shifts that would send walls of flames kilometres long into towns along the central ridge of the Blue

Mountains, places where hundreds of homes had been destroyed by fires in the past. This would be the last major firefighting operation overseen by Phil as RFS Commissioner. Soon after, he would resign from the RFS and be elected to parliament as the local state member for the Blue Mountains, later becoming NSW Minister for the Environment.

Prior to Black Summer, Victoria was the state in Australia most affected by bushfires and on the front line of increasing bushfire risk. It has been compared to California and France in terms of the volatility of the environment and regularity of major fires. The largest losses from bushfires occurred in 1851, 1898, 1939 and 1983, with the most damaging fire weather and fires happening about forty to fifty years apart.

In October 2008, well before the normal commencement of serious fire weather in Victoria, the Country Fire Authority warned that the bushfire season was shaping up badly, with extremely dry conditions following a drought. An El Niño increased the likelihood that Victoria would remain dry and that temperatures would be above normal. Authorities were already dealing with a larger number of fires than usual, but the hottest, driest, windiest weather would likely arrive early in the new year, with February historically producing the worst fire weather.

In the last week of January 2009, Victoria experienced an unprecedented heatwave, with temperatures in Melbourne exceeding 43°C for three days in a row, something that had never happened before. The forecast for Saturday 7 February was foreboding: temperatures in the 40s, accompanied by strong winds. Authorities warned that it would be a bad day for bushfires, and as the day approached, fire services and the state premier warned people to prepare for the worst. Through AFAC, a number of discussions were held with Victorian fire chiefs about the worsening conditions, and interstate fire services started to prepare for the likelihood that Victoria would need to seek urgent firefighting assistance.

On that day the temperature in Melbourne climbed to just under 47°C, accompanied by gale-force winds reaching storm force in some locations. I remember the day vividly, as Erris and Kate were visiting Melbourne that weekend. They later told me how incredibly hot it became: their eyes dried out and stung as heat radiated up from the footpaths. They went to the famous Queen Victoria Markets in the morning but were confronted with mayhem as gale-force winds tore through the sheltered but open areas, and the temperature surpassed 40°C before lunchtime. Safety concerns about flying debris and high temperatures forced authorities to close the markets. Erris and Kate sought refuge in an air-conditioned cinema and watched a movie to escape the blistering heat.

I was driving to the town of Corowa on the New South Wales / Victorian border to attend the centenary celebration of the local fire brigade. As my colleague Bill and I drove past parched paddocks, we were engulfed by a dust storm and had to pull off the road, unable to see more than about 5 metres ahead. The outside temperature was showing as 49°C. On arrival at the motel, an unusual sight greeted us: about a dozen people were standing, fully clothed and wearing hats, in the small swimming pool. Bill and I soon joined them, realising that in the blistering heat it was foolhardy to be exposed to the sun or even to take off your shoes to walk to the pool, as the ground was scorchingly hot. A thermometer said the water temperature was 31°C, but compared to the brutal air temperature it felt like ice.

Concerned about what might be unfolding in Victoria, I called Shane Fitzsimmons, now the NSW RFS commissioner. We both knew from news reports that Victoria was experiencing fire problems, but it was unclear at that stage just how bad things were. Shane and I decided to get ahead of the game and started to put together strike teams, groups of four or five fire trucks each with a separate commander, and moved them towards the border. Shane was able to get through to the CFA chief officer, Russell Rees, and I managed to talk to the CEO, Neil Bibby.

As is often the case in a fast-moving disaster, the picture in the central control centre in Melbourne was unclear. Fire controllers knew that there were many major bushfires, and unprecedented dangerous weather conditions, but little beyond that. Due to the sheer number of fires, decentralised command and control, power and communication outages caused by extreme weather, and strong winds preventing aircraft from flying on reconnaissance missions, they were experiencing great difficulty in piecing together a detailed view of the statewide situation. I felt for them because I could imagine being in exactly the same position, feeling powerless and bracing for inevitable bad news. Our strike teams started to make their way into Victoria in the early evening, but by then most of the damage had already been done.

Days later, Shane and I flew to Melbourne to provide a bit of moral support to the shocked, demoralised command team. I know that many of them continue to struggle to come to terms with what happened on what became known as Black Saturday, and with their feelings of powerlessness as they faced fires that could not be controlled by human efforts. Extreme weather drove flame fronts that in some locations reached average intensities of 88,000 kilowatts per metre (kW/m) of flame front – akin to a nuclear explosion or cramming 44,000 bar heaters into a one metre space – and drove towering convection columns 10–12 kilometres high into the stratosphere, bringing dry air and extreme winds down to the surface. Fires were so hot they generated storms that deposited burning embers up to 33 kilometres away, starting new fires.[3] A wind change after 5:30 pm turned long fire flanks into raging flame fronts (one 55 kilometres long) overwhelming firefighting efforts, destroying more buildings and causing more deaths.[4]

By late evening, the true scale of the disaster started to emerge. The first fatalities had been confirmed, but immediate access to some of the worst-hit areas was impossible due to ongoing fires, fallen trees, blocked roads and burnt bridges. It was feared, with good reason, that many more lives may have been lost in inaccessible areas.

From Coleraine in the west of the state to Churchill in the south-east, stories were emerging of firefighters forced to abandon any hope of fighting fires or saving property, concentrating solely on their own survival and saving as many other lives as possible.

The fire at Kinglake killed 159 people and destroyed hundreds of buildings when the super-heated smoke plume coupled with the upper atmosphere and formed a pyroconvective storm.[5] Like the earlier disaster in Canberra, a number of fires that day became plume-driven, bringing strong winds and dry air from the upper atmosphere down to the surface, making fires spread in multiple directions, burn extremely quickly and intensely, and generate embers that started new fires kilometres ahead of the main fire fronts.

More than 300 fires were reported throughout Victoria on Black Saturday, and at least fifteen of them raged out of control, destroying property and taking lives. The ultimate toll would be 2029 homes destroyed, thousands of other buildings damaged and destroyed, and 173 people killed, by far the worst fire disaster in terms of loss of life in Australia's history. The environmental damage to flora and fauna was immense, with vast areas denuded of all vegetation by fires that would have killed everything in their path.

The previous most recent large-loss fire in Victoria had been in 1983 when the Ash Wednesday fires destroyed nearly 2000 homes and killed forty-seven people. Over the preceding 150 years or so, the average return period for the drought and serious fire weather conditions that drove Victoria's worst fire disasters was forty-four years. Black Saturday happened just twenty-six years after Ash Wednesday, though my mathematician son Phil tells me that this is not necessarily 'statistically relevant', and may prove to be an outlier, therefore care needs to be taken in drawing any conclusions from this. However, it is certainly cause for concern.

The 2003 Canberra firestorm, the 2009 Black Saturday fires and the Royal Commission that looked into it resulted in the most fundamental changes to Australian firefighting doctrine since the Stretton Royal Commission following the 1939 bushfires.

The 1939 Bushfires Royal Commission resulted in more impetus being put into developing volunteer bushfire brigades; hazard reduction burning as a mitigation tool became more prominent; and fire authorities started to develop a policy on evacuation, colloquially called the 'stay or go' policy. This was based on the premise that bushfires are survivable and that well-prepared homes can be defended and saved if people stay with them. This view was reinforced following the Ash Wednesday fires in 1983, with a subsequent study finding that houses could be saved if people stayed behind to defend them.[6]

Findings of the Black Saturday Royal Commission resulted in a fundamental reset of national fire prevention, mitigation and firefighting doctrine. The loss of 173 lives, many of them *within* buildings, radically challenged paradigms, previous research findings and beliefs about bushfires, bushfire behaviour, and survivability.

AFAC, the peak council for fire and emergency services in Australia and New Zealand, coordinated national discussions that explicitly shifted the Prepare, Stay and Defend or Leave Early ('stay or go') policy to an overt focus on 'primacy of life'. Fire services had previously been reluctant to order mass evacuations, but under new policies, evacuations post-2009 became far more common.

Other fundamental changes followed: research conducted following the Canberra fires in 2003 and reinforced by the 2009 experience showed that off-the-scale fire dangers (as measured by the Forest Fire Danger Index – FFDI; see Appendix) were no longer rare, and that when the Index exceeded the maximum value of 100, lives and buildings were likely to be lost. Long-established fire danger ratings were revised: the former rating of Extreme (FFDI of 50–100) was subdivided into Severe (50–75) and Extreme (75–100). A new rating of Catastrophic / Code Red (over 100) was introduced. On days of forecast Catastrophic Fire Danger, fire services agreed nationally to change public messaging to recommend evacuation from high-risk areas the day before, warning that lives would likely be lost.

The Royal Commission thanked and congratulated firefighters, but criticised fire services for not disseminating current fire information, warnings and advice to the public. A new national hierarchy of warnings distinct from the FFDI was introduced, as well as technologies to immediately send the emergency warnings to mobile and fixed line phones in areas under threat. This was notable, as the previous focus of fire agencies had been mainly on fire prevention and mitigation (for example hazard reduction burning), and fire suppression. The changes were an acknowledgement that on some days firefighters would be almost powerless, because of extreme weather conditions, but needed to convey as much information as possible to the public.

The lessons from Canberra and Black Saturday were many, but some of the key take-outs were:

- There has been an increase in the frequency of serious fires since the late 1990s associated with more frequent extreme weather events.
- Climate change is increasing the frequency of extreme weather events, and therefore increasing fire risk.
- Fire seasons are lengthening in many parts of Australia.
- The frequency of seasons with very serious fire weather in Victoria may be increasing, meaning there will be more days when traditional mitigation tools such as hazard reduction burning are less effective than in the past.
- Long-term warming, higher evaporation rates and a reduction in rainfall in many parts of Australia are collectively drying out vegetation, leading to more intense fires due to an increase in the volume of fuel available to burn.
- The once-rare phenomenon of pyroconvective activity (fire-generated storms) that can lead to fires incapable of being controlled, may be increasing. This may be due to a combination of more extreme fire weather events combined with drier fuels that combust more readily and intensely.

Aerial firefighting also changed after the 2009 fires. The Victorian and New South Wales governments each trialled large and very large aerial fire tankers that could drop between 15,000 and 35,000 litres of water or fire retardant at a time. Long opposed by small sections of the firefighting industry due to concerns about cost effectiveness, use of large and very large aerial water bombers is now routine across Australia. They are expensive and not effective in all circumstances, but have proven to be a valuable tool in the fight against worsening fires as they can help contain small fires when they first start, slow down large fires and reduce fire intensity near buildings, giving firefighters on the ground a vital edge.

The cost effectiveness argument is curious, particularly as some of its proponents claim with little or no proof that expenditure on aircraft diverts funding away from prescribed burning. Nobody has ever claimed that aircraft are a panacea or that they can put out fires on their own but, as explained above, they can sometimes give a vital edge in the saving of life and property because timely, well-placed aerial drops can slow down or reduce the intensity of a fast-moving fire, enabling ground crews to get closer to the flames with their hoses.

Cost effectiveness was not raised during the response to the COVID-19 global pandemic despite Australia and the rest of the world plunging into recession; it was accepted that during a crisis, everything that could be done, had to be done, regardless of cost. During wartime, the cost of the military and defence efforts are rarely criticised. It should be the same when it comes to trying to adapt to the impacts of out of control climate change.

I applaud the New South Wales and Victorian authorities who had the vision to introduce larger firefighting aircraft to Australia following the 2009 fires, but it needs to be understood that they will not change the odds in favour of firefighters on the worst fire weather days such as those experienced in 2003, 2009 and 2019–20. As is the case with hazard reduction burning, aerial fire attack becomes less effective as fire weather worsens, and aircraft are unable to fly in strong winds. Nevertheless, they can give a vital edge in some circumstances. Proven, rapid fast-attack strategies using suitable aircraft need to be adequately funded and extended in Australia, as climate change forces us to adapt our response to fighting bushfires that are becoming bigger, burning hotter, and becoming increasingly more difficult to control and exponentially more destructive.

The ferocity of the weather conditions in 2009, the nature of the fires that they drove, and the scale of life lost were worse than anything experienced in the past. There was a lot of discussion in firefighting, research, and media circles about whether the fires might be associated with climate change. I was interviewed at the time by the *Sydney Morning Herald* and discussed climate change, how it was impacting bushfires, and how worsening fires were becoming harder for fire services to deal with despite better funding, equipment and technologies.

I was well within my area of expertise and knowledge. As an invited speaker I had delivered a paper at the Institution of Fire Engineers' International Fire Science Conference in Ireland in 2004 on the impacts of climate change on fire services, focusing particularly on weather and bushfires. Since 2007 I had been a member of the New South Wales Government's Climate Change Council, and as acting chair of the NSW State Emergency Management Committee in 2006 I had formed a climate change adaptation committee. Nothing that I said in the article was controversial or unsupported by science and observation as far as I was concerned.

But like other senior public servants forced to self-censor, I quickly learned that speaking publicly about climate change was out of bounds. I was told by a senior politician in no uncertain terms to keep out of the climate change debate and stick to fighting fires. This reinforced to me that, unlike the UK, where the science of climate change is accepted by all sides of politics, in Australia it has been turned into a political battleground, potentially making all of us casualties.

Tasmania has a long history of fires, but historically they had been infrequent, the worst fire weather events happening about 30 years apart since the late 1800s when record-keeping began. Much of the island is covered in wet rainforest and wet alpine plant communities formerly considered too wet to burn. The worst fires were in 1967 when around 1,500 homes in Hobart were destroyed and sixty-seven people were killed. Large fires didn't return until the late 1990s, but since then the frequency of serious fire weather, coupled with drought, has resulted in more fires, sometimes in areas with no previous fire history.

After a dry 2012, Tasmania experienced unexpected serious fire weather periodically from November 2012 up until the end of April 2013. During November and December 2012, serious fires broke out at Forcett, where several firefighters were injured after a sudden wind change; in the Central Lakes region; in Hobart suburbs, where vehicles and shacks were destroyed; and at Rhyndaston.

A significant heatwave was forecast for the first week of January 2013, affecting large parts of south-eastern Australia. On 4 January, Hobart recorded its highest temperature on record, 41.8°C, accompanied by Catastrophic Fire Danger indices at a number of locations, the first time that the new off-the-scale fire danger rating, introduced after Black Saturday in 2009, had been declared in Tasmania.

Fires broke out at numerous locations, stretching limited firefighting resources. In the town of Dunalley, sixty-five buildings were destroyed, including homes, the police station and shops. Many more homes were lost at other locations, a total of around 200, and nearly 50,000 hectares of bushland burned. Anyone who has travelled to Tasmania knows that it is significantly cooler than the mainland. Many locals seem to consider temperatures in the high 20s to be 'hot'. Temperatures in the high 30s, let alone the low 40s, are rare and unwelcome.

That word again – unprecedented – accurately describes the changed weather patterns leading to more frequent fires and the much higher temperatures being experienced in Tasmania. A prolonged drought eventually resulted in power shortages due to the failure of hydro-electric power generators, and rainforest and alpine vegetation formerly considered to be fireproof burned freely.

After very dry conditions through 2012 and 2013, September 2013 was the warmest September on record in about 75 per cent of New South Wales. Strong winds throughout the month resulted in widespread early bushfire activity, particularly on 10 and 26 September. On the 26th, a factory was lost to a grass fire in western Sydney: something highly unusual at that time of year because the weather conditions should not have been serious enough to support such intense burning. I attended that fire and was deeply concerned at the dryness, the hot and windy conditions so early in the season, and the ferocity of what should have been a routine blaze given the time of year. I watched as it burned as though we were in the midst of high summer.

Major fires were again starting early, before the official start of the 'normal' fire season. This was becoming almost routine, with many local councils and the RFS bringing forward the statutory bushfire danger period and fire restrictions by weeks and months almost every year.

The pattern of hot, dry, windy days leading into October 2013 would later be termed 'significant drying events', priming bushland fuels for major fires. Satellite imagery of the lower Blue Mountains showed that fuels were critically dry and tree canopies were stressed – a feature usually only associated with drought, but not on this occasion.

Professor Andy Pitman, Director of the ARC Centre of Excellence for Climate Extremes, later explained to me how the very warm winter in 2013 may have resulted in trees in the Blue Mountains, which in winter would normally go into a semi-dormant state, continuing to grow and draw water up through their root systems. This would have dried out soils at surface levels where root systems of shrubs and grasses were trying to obtain moisture, resulting in those fuels drying out and becoming highly flammable.

The overall effect would be soil dryness and fuel moisture levels following the warm, dry winter being far worse than suggested by drought index calculations based on rainfall, temperature and humidity. The significant heat events in September, coupled with days of drying winds, further primed the lower mountains for a fire catastrophe. This reinforced to me that traditional means of measuring dryness and fire risk were not keeping pace with changes in the environment driven by the warming climate.

On 10 October, temperatures in the Blue Mountains reached 35°C, with strong, gusty winds. Severe Bushfire Danger was experienced on that day in several New South Wales weather districts.

Three days later, significant fires broke out in the Hunter region north of Sydney, with six homes destroyed at Salt Ash near Newcastle.

On 16 October, a fire broke out at Marrangaroo, a large military reserve near Lithgow on the western side of the Blue Mountains. Local firefighters were unable to immediately access the mountainous area due to the danger of unexploded ordinance, and the fire quickly spread uphill into inaccessible country. With a day of significant fire weather forecast the next day, fire services knew that the fire would run hard to the east, but at that stage we had no idea just how hard it would burn and how much territory it would cover. It became known as the State Mine fire, and would burn for many days, eventually reaching western parts of Sydney in the Hawkesbury region about 80 km to the east.

Early on 17 October, as predicted, the State Mine fire intensified and started to spread rapidly under the influence of high temperatures, low humidity and strong westerly winds. It was largely inaccessible, and firefighters concentrated efforts on property protection on the fire's southern flanks. Fire intensity was too great to contemplate any form of direct attack on the fast-moving fire front that was spotting freely. Wilderness areas to the north were inaccessible under current fire conditions and air attack on such a large, intense fire front was futile except to assist ground crews with property protection. I was located at the State Operations Centre of the RFS with Commissioner Fitzsimmons, and once again I experienced the now familiar feeling of helplessness against the elements.

A new fire broke out in the upper Blue Mountains at Mount Victoria to the south of the State Mine fire. Local firefighters together with reinforcements from Sydney we had earlier staged in the area immediately went into property protection mode, with several properties destroyed and damaged as the fire jumped the Darling Causeway into the Grose Valley, an area that had burned just seven years earlier.

I vividly recall that day. At one stage I walked outside for a breath of fresh air to ease the tension. On walking back inside, a large screen in the Fire & Rescue NSW cubicle was showing live aerial footage of a crowning fire impacting on properties.

I assumed that it was the Mount Victoria fire, but was confused because there were so many homes and the terrain looked unfamiliar. There was a lot of activity in the huge operations room and Commissioner Fitzsimmons urgently waved me over. I asked what was going on.

'That's Winmalee,' he said. 'We're losing houses already. What can you send? We're throwing everything we can at it, but a lot of the staged units in the mountains are at Mount Victoria, so there's a lead-time before we can get out-of-area units there.'

I realised that this fire had the potential to destroy many homes, and to take lives. I called Assistant Commissioner Mark Brown, Fire & Rescue's Director of State Operations, and told him to start dispatching additional fire crews from Sydney and to arrange basic response coverage with what was left over. Mark pointed out that this would leave very little in reserve if another fire broke out, because a large number of RFS tankers from throughout Sydney were also being dispatched 'up the hill'.

This was a classic situation of damned if you do, damned if you don't. The likelihood of major property and life loss at Winmalee was clear and present. There is an old adage in the fire service – fight the fire you have, not the one you might have – meaning you should try to hit new fires as hard as you possibly can, rather than keeping resources in reserve, just in case. Either way, things can end badly.

The whole of Sydney and surrounding regions were sweltering under high temperatures and being swept by strong, gusty winds. It was quite likely that new fires would break out, as there were still hours of sunlight, high temperatures and high winds left before dark, when conditions normally moderate. Mark's concerns were spot on.

'Send whatever we have,' I said, 'and see how many trucks Logistics can scrape together from the workshops. Start recalling off-duty firefighters, put them on the reserve trucks and they can cover Sydney. We need to send everything we can. Right now,

Winmalee is ground zero – let me know straight awa,
fires start.'

Fire & Rescue NSW sent about ninety fire engines and 400 h.
fighters in the initial response, then followed this up with another
wave of fire trucks from country fire stations west of the Dividing
Range, which took hours to arrive. I wondered whether I would
be explaining my decision to the state coroner at some stage, and
hoped against hope that no further fires would break out.

But of course they did. At Balmoral in the Southern Highlands,
and at Munmorah on the Central Coast. More fires broke out in
Newcastle suburbs. Luckily, because of the distances involved,
we had not taken many resources away from Newcastle and the
Central Coast. Together with local RFS brigades, they were able
to protect most of the properties that were impacted. We sent
fire crews from the city of Wollongong to assist the RFS at the
Southern Highlands fires.

Word came back from Winmalee that senior officers in their
command cars and many of the first arriving Fire & Rescue crews
were not fighting fires – instead, by necessity, they were driving into
areas of heavy fire impact, running to houses, bundling people and
pets into the backs of cars and fire engines, and delivering them
to a local football oval where they were relatively safe. Firefighters
were in rescue mode for more than an hour and the events of
that day would see a number of officers medically retired because
of post-traumatic stress disorder. In just a few hours, more than
200 homes were destroyed by fire, most of them in the Linksview
fire at Winmalee.

Over the coming days, NSW would call on support from fire
services nationally in preparation for another day of Extreme
fire weather forecasted for Wednesday 23 October. Fire modelling suggested that the confluence of three major fires in the Blue

Mountains would amplify the forecasted fire weather, particularly the winds, and result in fires merging and impacting possibly thousands of homes. The 22nd was another sleepless night.

Thankfully, a thunderstorm in the Blue Mountains early that morning resulted in some heavy falls of rain, reducing the threat, but not extinguishing the fire. Reinforced by firefighters from South Australia, Victoria, Queensland, the Australian Capital Territory and Tasmania, we gradually gained the upper hand over the following week.

The 2013 fires were a further wake-up call, suggesting that well-established norms of climate and fire-weather interactions were changing. There had been some significant fires in the past on single days of serious fire weather, but in the absence of El Niño they were generally one-off events that lasted for a day and didn't result in significant property losses.

The implications of the Blue Mountains fires were startling. No El Niño, many days of Very High Fire Danger and above experienced from early September – before the official start of the bushfire danger season. This was remarkable because, firstly, the October 2013 fires, the most destructive in the state's history up until then, occurred at a time of year that had never previously produced fire weather resulting in major property loss. Previous major property losses in New South Wales had occurred from mid-November to January.

The early start of the season and the severity of the early fire weather unnerved me. Historically, nearly all major property loss fires in eastern Australia had happened during El Niño years, which are typically hotter and drier than average (see Appendix). Not only was the onset of serious fire weather very early, the El Niño Southern Oscillation (ENSO; see Appendix) was neutral, and therefore wind and weather patterns leading to hotter, drier conditions were not being amplified by this long-established driver of serious fire weather. The Climate Council dubbed 2013 the Angry Summer, at that time the hottest ever recorded in Australia.

The ramifications of this were quite stunning. It potentially meant that the climate had warmed and weather patterns changed to a point that the super-charging effect of an El Niño may no longer be necessary for the East Coast to experience a damaging fire season. My other big concern was, if this proved to be correct, what would an El Niño-driven fire season look like in the future? The 2013 New South Wales bushfires destroyed 280 homes (fifty-three near Coonabarabran in January, when a pyroconvective storm resulted in extreme fire behaviour[7], and then 227 in October), and the Tasmanian fires in January destroyed around 200 homes, making 2013 a particularly damaging year. Tasmania had experienced the worst fire danger for many years and the highest property losses since the 1967 fires that tore into the suburbs of Hobart destroying more than 1,500 homes and killing sixty-seven people. Even back in 2013, and despite the severity of the fires, any suggestion that New South Wales would soon experience property losses on a scale dwarfing those of previous catastrophic fires in Tasmania and Victoria would have been scoffed at, because we had never experienced the confluence of drought, weather and fuel conditions capable of producing such firestorms.

6

CANARIES IN THE COAL MINE

IN THE YEARS FOLLOWING the 2013 fires, I saw how areas prone to bushfire continued to exhibit extremes of fire weather at times and in places they had not before, making the task of predicting, avoiding, and fighting bushfires increasingly difficult. I watched with trepidation as reports of increasingly intense fires and increasing property and life loss rolled in across the country.

In January 2017 I retired as Commissioner of Fire & Rescue NSW after just under thirty-nine years of service. Just prior to this I rejoined the volunteer brigade where Dad was still a life member and where he and I had fought many fires together in the past. I felt that I had come full circle, and reflected on decades of worsening fires. Dad still had a keen interest in what was happening to the bush and to the climate. We talked for hours about the changing environment, agreeing that something bad was coming our way.

It wasn't long before I was out among fires again, but in a totally different context: directly on the front line, again dragging hoses through the bush chasing flame fronts rather than sitting in the State Operations Centre trying to sort out competing resource demands and preparing briefings for politicians. I loved being back out in the bush.

In early 2017, New South Wales fire services were becoming apprehensive about the possibility of serious fires despite the absence of El Niño, previously an almost essential ingredient for a serious fire season to occur. In 2016 there had been significant spring rains up until October in western parts of the state, resulting in prolific growth of grass and scrub. Farmers were happy as this provided them with feed for their livestock and good growing conditions for crops following a difficult period of drought. But among the firefighting community, it was worrying.

Farmers know that better growing conditions can be a double-edged sword – leading to better harvests and more stockfeed, but also increased risk of bushfires later in the season. As the year progressed, temperatures increased and grassland fuels started to dry and cure (see Appendix), ripe for a bushfire.

From the Upper Hunter region to the Liverpool Plains and Gunnedah to the west, locals were becoming increasingly concerned about the emerging fire risk. By February 2017, grassland fuels were mostly 100 per cent cured – dried out and ready to burn, needing only a spark and suitable weather conditions for a major fire to break out.

Leading up to the Sir Ivan fire in New South Wales, January and February 2017 had seen temperatures well above average in many areas of the state. The summer of 2016–17 broke more heat records, surpassing 2013 as the hottest in history.[8] In early February, many locations recorded temperatures above 40°C and very low humidity levels. The combined effects of dryness and high temperatures dried out forests and grasslands, priming them for conflagration.

Extreme Bushfire Danger was forecast for Saturday 11 February in many parts of New South Wales, with forecast temperatures of 42°C, relative humidity of 10 per cent and winds from the west / north-west at 45 kilometres per hour, gusting to 70 kilometres per hour in the Castlereagh region. Even worse conditions were forecast for the following day with RFS Commissioner Fitzsimmons later telling a Coronial Inquiry that they were the worst fire weather conditions ever recorded in New South Wales.[9] These conditions would later be surpassed by large margins on multiple occasions during the record-breaking Black Summer of 2019–20 under three years later.

Around midday on Saturday 11 February, smoke was seen coming from an unoccupied farming property near the town of Dunedoo. A smouldering fence post struck by lightning in an electrical storm eight days earlier was later identified as the likely ignition point.

Local RFS units and farmers responded and worked into the night, but were hampered by high fuel loads, erratic winds and difficult terrain. By morning, the fire had not been contained and continued to burn in rugged, inaccessible areas. Catastrophic Fire Danger was forecast for Sunday 12 February across the fire area.

At around 10 am, the feared weather conditions arrived. Very strong winds accompanied by high temperatures and very low relative humidity drove the fire through scrub and forest, with numerous spot fires starting up to 500 metres ahead of the main fire. The spot fires rapidly grew, then themselves spawned even more spot fires, meaning that the rate of spread was phenomenal. Firefighters reported hearing the roar of the wind as the fire approached, and some said that water jets from fire hoses were ineffective against the walls of flame driven at them by hot winds.

By 11:30 am emergency warnings were being broadcast on mobile and fixed-line telephones advising people to evacuate if possible or to urgently seek shelter. Orders were given for firefighters to withdraw from fire fronts and to seek shelter, as the fire raged through tree canopies and the fire grew at the incredible rate of 6,000 hectares per hour (imagine 6,000 rugby fields an hour, or 100 rugby fields every minute).

Later in the afternoon, after 5 pm, a fire-generated storm formed above the fire. The massive convection column from the fire created a storm cloud as it pushed high into the stratosphere, but it didn't produce any useful rain. Lightning from the storm cloud was recorded up to 100 kilometres away.

Thankfully, the Sir Ivan firestorm coincided with a strong wind change, causing the storm cloud to de-couple, or separate, from the convection column. Experts say that had this not occurred there almost certainly would have been substantially more property loss, the fire would have spread over a much larger area and burned more intensely, and lives could have been lost.

Ultimately, the Sir Ivan fire burned more than 55,000 hectares, destroyed thirty-five homes, many farm buildings and farm machinery.[10] In hindsight, the fire and the frightening weather conditions that spawned it were a foretaste of things to come. Dr Simon Heemstra from the NSW RFS clearly identified a hotter atmosphere leading to an extended fire season as a reason for the fires, explained the scientific evidence of changing fire weather, and told the inquiry into the fire, 'The climate is changing and we are getting more extreme fire behaviour as a result as the atmosphere is hotter and more unstable.'[11]

In early 2018, my best mate, mentor and hero had some minor surgery but never recovered. Dad remained stoic to the end. Together with my sisters and Erris, we made sure that he was able to stay at home, just as we had with Mum when she became ill and passed away in 2008. Erris's nursing skills and the care provided by Kim and Rob made it all possible. He passed away one afternoon in

March with the four of us around him, holding him as he took his last peaceful breath. Dad was ninety-three, and the last outing he'd gone to was the local RFS brigade Christmas party. The captain at the time, Todd, knowing I was interstate, had sent a four-wheel drive to pick Dad up, then dropped him back home afterwards. He had a great time and told me all about it when I returned – a great memory.

His memorial service was held in the community hall that he had helped build, and a local brigade brought along their fully restored 1942 Blitz bushfire tanker, just like the one Dad had often worked on, and parked it outside. As the packed memorial started, there were three long blasts on the old fire siren that hadn't been used since pagers were introduced. It was fitting that locals should know that one of their long-term protectors had hung up his helmet for the last time. Dad, not a religious man, realised weeks before that he wasn't going to recover. He had joked that he would see a lot of old friends 'down below'; but if there is a heaven and a hell, he definitely went up, not down, and he would have met up with Mum and Terry again.

On the evening of the day Dad died, a huge orange full moon rose over the ocean. Mum had always loved a full moon, and we would often drive to the beach and sit on a headland to watch it when we were kids. We like to think that Mum was sending us a signal that Dad was now back with her and in good hands.

I felt a huge sense of emptiness when we lost Dad, just as I had when my beautiful mum passed away. Since my retirement we had travelled together, and we would catch up most days for his famous coffee and a game of Yahtzee. But regardless of how I felt, the worsening weather and fires didn't miss a beat: 2018 was going to be another bad year. I would have to battle on without the benefit of Dad's insights and wisdom.

*

Tathra is a popular tourist destination on the picturesque south coast of New South Wales, home to many retirees, with holiday homes often rented out during the holiday season.

The south coast has a long history of major fires, with dense mainly wet sclerophyll forests stretching inland to alpine areas and all the way to Canberra. In 1968, major fires swept down from the mountains and caused significant damage to the then relatively sparsely populated region. Since then a number of other fires have burned large tracts of land.

Meteorologists have noted a long-term drying of the region, with up to 12 per cent less winter rainfall over a twenty-year period.[12] During a conversation with Dr Heemstra back in 2018, I recall him expressing deep concerns about the South Coast, explaining how the decades-long drying trend associated with climate change had primed the area for major fires and the possibility of major loss of life and property, exacerbated by the expansion of communities in the years since the last major fires. Unfortunately, his prediction came true.

Based on previous experience, by late March, the fire season should have already finished in this region, with temperatures decreasing and fire danger easing. However, on Sunday 18 March 2018, the New South Wales far south coast experienced unseasonal extreme fire weather. At 1 pm the temperature reached 37°C, the relative humidity dropped to 17 per cent and north-westerly winds were blowing at 44 kilometres per hour, gusting to 72 kilometres per hour.[13] A maximum temperature of 38.4°C was recorded that afternoon and the Bureau of Meteorology later reported that the extreme weather conditions resulted in one of the highest fire danger index readings ever recorded there, despite it being past the normal peak fire danger season. It was by far the worst fire weather ever experienced in autumn, when fire danger would normally have been moderating as the weather cooled.

At about 11:15 am the first emergency call reporting a bushfire was received, followed by further calls at about 12:28 pm that another fire had broken out near Reedy Swamp, west of Tathra. The fire would later jump the Bega River and spread unimpeded into the Tathra township, destroying sixty-nine houses, thirty caravans and cabins, and damaging another thirty-nine homes.

The statutory bushfire danger period in New South Wales ends on 31 March each year because historically, by this time of year, it is very unusual to experience serious fire weather anywhere in the state as autumn heralds lower temperatures.

The Bureau of Meteorology later reported that April 2018 was the driest April since 1997 and the fourth driest ever recorded in south-east Australia.[14] The month was 'exceptionally warm' and nationally the second hottest April on record. The Bureau of Meteorology reported that the heat was, 'unprecedented in many areas in April for its intensity, its persistence or both.'[15]

With the Tathra fires barely extinguished, the combination of dryness and heat set the scene for further fires, despite it being outside the period generally recognised as problematic for bush-fires. Sydney set an all-time temperature record for April of nine days above 25°C, with several days exceeding 34°C, and the hottest day ever recorded in April experienced on 9 April when it reached 35.4°C. Hot weather then set in from 12 to 14 April, together with strong north-westerly winds, unusual for this time of year, accompanied by low relative humidity.[16]

On Saturday 14 April, during Severe Fire Weather conditions, a fire broke out in grass and scrub near Casula railway station in Sydney's south-west. It appeared to have been deliberately lit. Under the influence of high winds, it spread quickly through parched grass and scrub and then eucalypt forest, over the next two days impacting heavily on populated suburbs including Wattle Grove, Alfords Point, Voyager Point, the Holsworthy Barracks, Menai, Barden Ridge, Pleasure Point, Picnic Point, Illawong and Bangor.

Through the efforts of hundreds of firefighters from Fire & Rescue NSW and the RFS, hundreds of homes and other buildings were saved. Another large fire would burn through the military reserve at Holsworthy in July 2018, three months before the start of the official fire danger season.

Unseasonal conditions of Very High Fire Danger along the entire New South Wales coast returned on 15 August 2018. After a dry winter, strong westerly winds gusting up to 95 kilometres per hour were experienced in some areas, together with unseasonably high temperatures. Many fires broke out and burned out of control. Emergency warnings were issued early in the morning on the south coast west of Milton and later in the Bega Valley, where a home was destroyed. Fires also broke out at Port Stephens – north of Newcastle.

By 19 August there were dozens of fires burning along the coastal strip in New South Wales, from the Queensland to the Victorian borders, and on that day multiple emergency warnings were issued by the RFS. I deployed to a number of fires, including fires burning near Port Stephens north of Newcastle.

Fires continued to burn through September with Severe Fire Danger experienced over a wide area on September 15, driven by low humidity and strong winds. Serious fire weather continued through October with all the portents of a long, serious fire season. Thankfully, rains arrived in November and the situation improved markedly, enabling New South Wales to send assistance to Queensland and Tasmania where fire conditions continued to deteriorate.

In mid-2018, Queensland fire authorities feared that a bad fire season was likely after yet another failure of winter rains. By early August, fires were already burning in Queensland.

According to Lee Johnson, former Commissioner of Queensland Fire and Emergency Services and board member of the Bushfire & Natural Hazards Cooperative Research Centre (BNHCRC), bushfires had not historically been considered a significant risk in Queensland, particularly when compared to the damage historically caused by floods and cyclones.

Bad fire seasons, fire intensity and their destructiveness in semi-tropical Queensland had historically been nothing like that experienced in southern states, and serious fire years were comparatively infrequent. Lee described to me how it was rare for Queensland to suffer fires with significant property losses or fires covering large, forested areas over an extended period. During his career of forty years he witnessed firsthand a gradual worsening of bushfire weather and a lengthening of fire seasons. Bushfire risk is now moving quickly up the scale of disaster threats in Queensland, particularly for peri-urban areas in central and south-east Queensland.

Lee had seen for himself, and had carefully documented, increasing impacts of weather systems exacerbated by climate change in Queensland since Tropical Cyclone Larry hit Innisfail in March 2006. One of the predicted effects of climate change is that there will be fewer cyclones overall, but when they occur, they are likely to be more intense, destructive and deadly than in the past because rising ocean surface temperatures and a warmer, wetter atmosphere provides a larger source of energy for tropical cyclones to draw on, once they are formed. Each year since 2006, Queensland has had to deal with at least one major natural hazard disaster and sometimes multiple, compounding events. Queensland's total losses from extreme weather disasters since the 1970s are around three times those of Victoria, and about 50 per cent greater than New South Wales. On a per-person basis, Queensland's losses were more than twice the national average.[17] Bushfires are not the only looming climate disaster threatening Australians.

The southern states and south-western parts of Western Australia had historically been far more prone to regular, large-scale bushfires and property loss. According to Lee, over the previous decade the bushfire situation in Queensland had become increasingly volatile because of a prolonged drying trend, rising temperatures and a significant increase in the number of days each year of serious fire danger during lengthening fire seasons.

A review by the Queensland Inspector-General of Emergency Management found that climate change is driving an escalation in the number and severity of heatwaves and of dryness, resulting in increased fire danger. A background paper prepared to inform the broader review cautioned that what once were considered to be 'out of scale' conditions will increasingly become part of normal variability in Queensland.[18]

Notably, the 2018 Queensland Fires burned in a variety of vegetation types, including sub-tropical rainforest that had never before burned intensely, with the 2018 review noting, 'Many Australian vegetation types with no previous history of burning are drying out and posing a greater risk.'[19] Ultimately, Queensland's 2018–19 bushfire season ended abruptly with record flooding rains in January 2019 that inundated homes near Townsville and drowned hundreds of thousands of cattle over huge areas. Queensland went from one extreme to another, an increasingly common feature of a warming climate seen in other parts of the world that ricochet from drought to flood with increasing regularity. New South Wales also experienced rainfall extremes in 2020 and 2021. Research has identified that for every 1°C increase in temperature, the atmosphere can carry 7 per cent more water vapour, increasing the probability of extreme downpours.[20]

At the same time, Tasmania, accustomed to major fire seasons occurring about once every thirty years, was still reeling from major fire seasons in 1997, 2006, 2013 and 2016.

In the summer of 2018, as had occurred just two years previously, dry lightning storms repeatedly swept across the state, starting dozens of fires. Interested to get a firsthand perspective, I spoke to former Tasmania Fire Service chief fire officers Mike Brown and John Gledhill who told me that they could not remember dry lightning storms like these during their decades of firefighting. They also said that the heavily forested World Heritage areas affected by the fires had historically been too wet to burn intensely and had not been considered particularly hazardous.

Tony Blanks, former head of fire control for the Tasmanian Parks and Wildlife Service and later for Forestry Tasmania (now Sustainable Timber Tasmania), described how Tasmania had undergone a long-term drying trend that had affected fuels in the High Country and in wet eucalypt and rainforests, increasing the likelihood of fires from lightning strikes. This accords with research into dry lightning storms and fire ignitions by lightning that established that higher temperatures, lower relative humidity, and drier fuels are associated with greater chance of ignition per lightning stroke.[21] Climate change is affecting all of these parameters, therefore making lightning-caused fires more likely.

Major fires in 2018 and 2019 covered the largest area burnt since the deadly Tasmanian fires of 1967. Some of the World Heritage Wilderness Areas that burned had previously been untouched by intense fire for millions of years. Sensitive alpine plant communities, including ancient trees like King Billy Pines and Pencil Pines, as well as mature temperate rainforests, were ignited by a series of dry lightning storms, particularly on 14 and 15 January when 2,400 lightning strikes were recorded, sparking seventy new fires.

Fire weather conditions leading up to December 2018 were not particularly serious and did not cause any significant concern. However, from mid-December until February there was little rain, with January recording only about 20 per cent of average rainfall for the state. Temperatures were above average and there was a rapid worsening of bushfire conditions.

A massive mobilisation of interstate support saw firefighters and equipment from across Australia sent to assist during the lengthy campaign. An independent review of the fires found that climate change is adversely impacting the length, frequency and severity of Tasmanian fire seasons.[22] The fact that Tasmania had experienced its fifth significant bushfire season since 1997 is notable, given the previous experience of decades between major fire seasons.

As 2018 drew to a close, there was growing discomfort in firefighting circles. Each year was getting hotter and drier, lengthening fire seasons were now a fact of life, and fire weather was intensifying – driving bigger, hotter, more destructive fires. The coming year was viewed with trepidation as a stubborn drought parched the landscape, coming on top of a twenty-year drying and heating trend.

We were witnessing climate change in action, no question about it; something even the most dedicated climate change deniers were having trouble explaining away. If significant rains didn't arrive in the winter of 2019, the east coast of Australia would be primed for an inferno.

There was broad agreement among the firefighting fraternity that New South Wales had dodged a bullet in 2018. The drought conditions, long-term drying, high temperatures and early onset of serious fire weather had stretched resources. Dodging a bullet is probably not a term that Queenslanders and Tasmanians would agree with, though, given the widespread fires and damage that they experienced.

When major fires started to break out in August 2018 in Queensland and northern New South Wales, there were no large air tankers or helicopters available because lease arrangements hadn't foreseen a need so early in the season. At this time, a bad Californian fire season that would kill nearly 100 people and destroy about 20,000 buildings was in full swing, and a contingent of senior fire managers from Australia and New Zealand had been sent to assist in the USA on the assumption that Australia would not be needing them. This proved to be wrong, reflecting the increasing overlap between southern and northern hemisphere fire seasons as well as increasing international competition for firefighting resources.

By early 2019 in the New South Wales Northern Tablelands, towns like Guyra were running out of drinking water, owing to serious drought exacerbated by higher temperatures which led to high rates of evaporation. Major fires broke out in early 2019, with bushfire emergencies declared in February in the Tenterfield and Inverell areas, then in March near the towns of Tenterfield and Glen Innes. Homes and other buildings were lost to the flames.

I had retired from Fire & Rescue NSW in January 2017 and was back at Terrey Hills RFS brigade where I had been a volunteer in the 1970s and early 80s, and where Dad had been a life member. I was becoming increasingly concerned about the inevitability of a serious fire season as the drought intensified.

Sydney had been spared major fires because of rain and storms from November 2018, but there was huge potential for fires in the north of the state early in 2019. When major fires broke out, reinforcements were sent by the RFS, National Parks and Wildlife Service, Forestry Corporation of NSW, and Fire & Rescue NSW, to assist local firefighters. I worked on two five-day deployments: at Tingha, outside the regional centre of Inverell, in February; then at Torrington Plateau, north-west of Glen Innes, in March. The area was tinder dry and fires were spreading freely even when winds were relatively mild.

The February deployment happened mid-week, with an urgent call for assistance after fires broke out and spread quickly near Tenterfield and Tabulam, not far from the Queensland border, destroying homes and other buildings. Major fires were also burning on the Queensland side of the border.

Strike teams were sent from Sydney and other parts of New South Wales to assist. I was placed in charge of one of them and our convoy of five fire trucks drove for six hours through the night from Sydney to Tamworth, the next morning setting off early for the Fire Control Centre at Glen Innes, two hours further north. Originally, the plan was to be briefed, fed, then sent to the fires surrounding Tenterfield and Tabulam where Severe Fire Danger was once again expected.

However, on arrival at the Fire Control Centre we were urgently reassigned to respond to a fast-moving blaze near Inverell to the west. The fire had broken out the previous evening and as weather conditions deteriorated, fire was threatening properties in the small town of Tingha, as well as surrounding isolated rural homes and farms.

As can often be the case in a rapidly developing situation, my team and I found it hard to establish contact with local command, as locals had been fighting the fire for more than twelve hours and the fire now covered a large area. It had raced through grassland, farmland and forest, destroying a number of properties, and it was difficult to establish where the fire perimeter and various fronts were.

As I drove through flames and dodged falling trees, we saw a massive black smoke plume ahead, which we discovered was a car and truck wrecking yard covering about one hectare. Trying to deal with that fire was futile, as petrol and LPG tanks were exploding into fireballs, throwing debris in all directions. We had to leave it to eventually burn itself out and instead concentrated on protecting properties that were considered saveable. There was no point in trying to contain and stop the main fire fronts that were burning quickly and intensely over a wide area.

As we headed back towards a small local airstrip, where a new front was threatening farms, a NSW Forests Corporation officer flagged us down and asked if we could stay to protect an isolated house and sheds on the main road. The property was surrounded by fire, but the wind direction meant that the flames were flanking, going past the property, and were therefore lower in height and intensity. I spoke to Matt Hunter, the very experienced officer in charge of one of the tankers, and he said that he was comfortable setting up to protect the house.

Almost immediately after Matt's tanker entered the driveway, there was a slight change in wind direction. The flames that had been about 2–3 metres high suddenly leapt into the treetops, 10–15 metres above us, and we were showered with burning embers and engulfed in thick smoke. Near the road, garden beds made from old car tyres burst into flames, with thick, black, oily smoke and flames blocking any further access to the property.

My offsider was Bill Dunlop, an experienced firefighter whose job was to drive, monitor radio messages, keep records and be a second set of eyes and ears, helping me to make critical decisions and to deploy the fire trucks under my command. We had to urgently relocate because the radiant heat through the windscreen of our four-wheel drive was unbearable, and trees were falling as the wind intensified. When I was unable to raise the tanker by radio, I sent an urgent 'Red' message by radio to the Fire Control Centre reporting a possible fire overrun, requesting urgent air support. I was hoping that a helicopter might be close by and be able to drop a load of water onto the tanker.

I feared for the lives of the crew and had a sense of dread. Bill and I felt utterly powerless. We could see that a number of sheds and the house were now well alight. Small explosions mushroomed through the dense black smoke. Just then, another one of our tankers arrived. Keith Davies, the officer in charge, had heard my radio message and raced back to find us. Without hesitation, the tanker and crew drove through the smoke and flames and then radioed the message I had hoped for, 'All accounted for and safe.' At that point a water-bombing helicopter arrived overhead.

I later learned that Keith and his crew arrived just as Matt's tanker ran out of water. Matt and his crew, after abandoning their hoses and running to the tanker for shelter, pulled down the aluminised window shades and huddled inside the cabin under protective woollen blankets. Caryn Thompson, the driver and a very experienced firefighter, activated protective water sprays that covered the outside of the tanker with a halo of cooling water. This probably saved their lives. It was a frightening experience for all of us, and it highlighted the dangers of firefighting and the strength of character of firefighters who willingly place themselves in harm's way.

The house and a number of sheds were destroyed. A hand-painted wooden For Sale sign at the front gate smouldered, and I remember the owner, an elderly man, standing silently across the road, obviously in shock. We discovered that the surrounding bushland was littered with dozens of old car and truck bodies and hundreds of truck and car tyres, generating thick, choking, black smoke and very intense heat and flames.

I stood down Matt's crew and went with them to a nearby RFS station for an initial debrief and a rest. They later asked if they could get back onto the front line. My admiration for all of them was immense, mainly because of the way they supported each other after the harrowing experience. It really had been a life or death situation, and if not for their experience, training and teamwork, it might have ended tragically. Keith and his crew had arrived at precisely the right moment, and the two tankers were able to drive out of the property to safety.

All of this had happened within the first 45 minutes of arrival, but thankfully the next few hours, while busy, were not nearly as eventful. We spent them chasing fire fronts and protecting properties – what we call 'hit and run' firefighting, not even attempting to engage the multiple fire fronts. As soon as a property was relatively safe and secure, we would move on to the next one.

Later that day, I received a phone call from the Fire Control Centre. They told me that there were reports of a missing person, possibly deceased, on a small farm. They didn't want to broadcast anything on the radio as the media might overhear and release details before relatives were contacted. Police couldn't reach the area because of the intense fires.

Bill and I made our way along a dirt road for several kilometres, again dodging fallen trees, and, after speaking to a local farmer, located the property. It was in a black and grey wasteland with smouldering trees, burning logs and tree stumps for as far as we could see. The smoke was thick so we both put on goggles, helmets and P2 particulate masks.

A shearing shed was well alight about 200 metres inside the fenceline, but a small farmhouse seemed to have escaped unscathed despite the fire burning within a few metres of it. There was an old car next to the dilapidated house, which was not a good sign – we wondered where the driver was.

Several farm dogs were chained up next to the house, cowering, panting and quiet – they must have been terrified as fires raged around them. None of them barked or even acknowledged we were there. We gave them some water to drink then set out on the unwelcome assignment to locate the driver. The informant had suggested that a person had been trapped in the shearing shed and not been heard from again as the fierce fire swept through the area.

I explained to Bill what to look for – or, more precisely, what to smell for. I knew from unfortunate experience that burning flesh has a characteristic smell, which I described to Bill. Sometimes a burning body can emit greenish- or blueish-coloured flames. I was pretty matter of fact when telling Bill this but realised by the look on his face that maybe I needed to work on my bedside manner. An unfortunate by-product of witnessing so much tragedy during my long career was becoming a bit hardened to people's sensibilities. Another legacy of my decades of service was learning late in my career that I was suffering from PTSD. While I could handle a lot, there were certain things that could also unexpectedly trigger me and play hell with my emotions: fellow firefighters being killed or injured on the job was one of my known triggers, so the previous situation with Matt and his crew had rattled me. Thankfully, on this occasion we found no evidence of a body and reported this back to Fire Control. Police later confirmed that the owner had escaped safely.

The next day I was assigned as division commander of a large section of the fire, under the command of the local incident controller, RFS volunteer Deputy Group Captain Ray White. Over a couple of days in the field, Ray and I became good friends. He did a remarkable job coordinating a major firefighting operation by working from the back of an old Toyota Troop Carrier set up with radios and parked in a desolate, rocky paddock on the fringes of Tingha.

Ray explained to me that it was a relatively elevated spot and provided fairly good radio coverage. His son, Alex, who had been co-opted as radio operator, was the senior deputy captain of his local RFS brigade. From there, they coordinated many units from the RFS, Fire & Rescue NSW, NSW Forestry Corporation, and a number of helicopters and small fixed-wing aircraft.

Ray is a self-employed tour guide and knew the local area like the back of his hand, the kind of no-nonsense bloke who was obviously very competent. He showed me a map of the area, indicated where he wanted me to go, and gave a general outline of what to expect. He warned me that there probably wouldn't be any radio or mobile phone coverage, which proved to be the case. I told him that I would probably 'Ask for forgiveness afterwards, not permission beforehand.' Ray gave me the thumbs-up. That worked for both of us.

Later that morning, after we had stopped a fire reaching a farmhouse and sheds, I pulled together a group of local farmers and farmhands who had arrived with firefighting equipment to help us. I explained that the fire front was irregular, and that bad fire weather was again on its way. The only way that I could see for us to stop the fire spreading was to conduct a backburn from a road that ran through the area and to use fire trails as control lines – a couple of kilometres of burning, with no guarantee of success. To do this safely, given that my division only had about eleven fire tankers and a couple of aircraft assigned, I would need the help of local farmers using their tanker trailers to secure fire edges behind us. This would release fire tankers and crews to leapfrog each other and keep ahead of the main fire while moving our burn along as quickly as we could.

The farmers all nodded their agreement and gave suggestions about access. I saw one of them, a gnarled-looking veteran of the bush, standing off to the side, with slumped shoulders, looking at the ground. I went over, and asked, 'Are you okay? Do you think we should try something else?'

I'll never forget his reply. He looked me straight in the eye and said, 'I know where you're coming from, mate, and what you've mapped out is the only bloody option. When you look at the grass out there, I know that you just see fire fuel. But what I see is winter feed for my stock. We've already been handfeeding for months and the dams and creeks are dry. My family has been on this land for over 100 years, but if we don't get rain before May, we'll have to walk away. Even if we get rain after that, it will be too cold for anything to grow.'

As it turned out, the much-needed autumn rains didn't arrive. For the third year in a row, there were no winter rains either. In 2019 the Northern Tablelands region received 77 per cent less rainfall than average, and towns ran out of drinking water.

After another dry autumn, the fires would return to that area with a vengeance in winter, three months before the official start of the fire season on 1 October. The Northern Tablelands region had the unwanted distinction of being ground zero for the start of the longest, most destructive fire season in Australia's recorded history. By October 2019, huge areas were on fire and properties had already been lost to the flames.

I often wonder how that farmer fared and I hope that he held on and stayed on the family property. It's him, and people like him, who live in close connection with the land and rely on it directly for their wellbeing, who will continue to carry a disproportionate burden from our changing climate, compared to the rest of us, as the situation continues to deteriorate. It's not the much-derided so-called greenies who are the real canaries in the coal mine of our mounting environmental catastrophe – it's our farmers and primary producers. These are the people who grow our food, make fibres for our clothes, and work to provide us with many aspects of our high quality of life. It's them in particular whom our government has abandoned by continuing to refuse to act on climate change.

After the early-season fires the year before in 2018, I had watched dryness extending across bushfire-prone parts of Australia, focused mainly on New South Wales because I live there and know it best. I felt a deep sense of unease when the Northern Tablelands fires in February and March 2019 were not followed by winter rains. I had seen firsthand the desperation of farmers unable to feed stock or grow crops, and small towns relying on truckloads of water each day just for the simple basics of drinking, cooking and flushing toilets. Fires started burning again in Queensland and New South Wales from July 2019. For months I had been speaking to other fire chiefs across Australia, serving and retired, and they all broadly shared similar concerns. There was a shared sense of foreboding.

2018 had enjoyed some moisture in late spring, which moderated the fire weather in New South Wales and then Queensland, but the 2019 seasonal outlook had no such indications. If the notoriously unstable conditions of late October and November produced no rain, storms and elevated humidity, we would be in serious trouble across a vast swath of the east and south-east of Australia, from southern Queensland to New South Wales and the

Australian Capital Territory, down into Victoria's East Gippsland region and probably also into South Australia. The seasonal outlook from the Bushfire and Natural Hazards Research Centre reinforced my concerns.

In early 2019, after returning from the Northern Tablelands fires, I started to call former fire chief colleagues to canvass their thoughts. I called former Queensland Fire and Emergency Services commissioner Lee Johnson; former chief officer of the Tasmania Fire Service Mike Brown; former Victorian Emergency Management commissioner Craig Lapsley; former head of firefighting for the NSW National Parks and Wildlife Service, Bob Conroy; former commissioner of the Western Australia Department of Fire and Emergency Services, Wayne Gregson; former deputy chief officer of the South Australian Country Fire Service, Craig Lawson; former commissioner of the NSW Rural Fire Service, Phil Koperberg; former chief officer of Forest Fire Management Victoria, Ewan Waller; former head of firefighting for Tasmanian forestry and national parks, Tony Blanks; former CEO of the Victorian Country Fire Authority, Neil Bibby; former chief fire control officer for the Northern Territory, Stephen Sutton; and former CEO of AFAC, Naomi Brown.

All of them said that they were deeply concerned, not just by how 2019 was shaping up, but by the changes in climate they had witnessed over decades, accompanied by worsening fire weather. Now unencumbered by government positions, each of them was becoming increasingly vocal about the same issues: longer fire seasons; a long-term reduction in winter rainfall; more days of serious fire weather; more intense temperatures and winds; more heatwaves driving fires. All of them were concerned about the increasing overlap of fire seasons between jurisdictions, making it more difficult to call for and send help.

I remember briefing the Climate Council as a relatively new member in March and April 2019 about my profound misgivings and anxiety about the upcoming fire season, and about what I was hearing from other experts across the country. The anxiety was not confined to Australia. Colleagues in the USA, where California had just endured its worst-ever fire season that saw the city of Paradise all-but obliterated, were also on the front line of climate-change-influenced extreme fire weather.

I floated the idea of a coalition of former fire and emergency service chiefs coming together to sound a warning about the upcoming fire season and give eyewitness accounts of how climate change was super-charging bushfire risks in Australia and around the world.

I described to the Climate Council how such a group could comment authoritatively in relation not only to bushfires, but also to other extreme weather events including worsening storms, floods, cyclones and heatwaves – natural disasters that were all showing clear climate-change signals. This would mean bringing on board not only former fire service, national parks and forestry fire chiefs, but also former heads of State Emergency Services, those agencies primarily tasked with response to flood and storm emergencies.

The CEO of the Climate Council, Amanda McKenzie; the head of research, Dr Martin Rice; and the communications director, Lisa Upton, talked me through what such a group might look like, how it might be formed, and teased out from me what we might be able to do to raise awareness of the impacts of climate change on natural disasters. Amanda and the Climate Council committed to assist me in setting up a group and providing media and administrative support.

Throughout this period, I had been talking to Erris about my concerns and also about my fears for the futures of our grandchildren as climate-driven natural disasters worsened. She supported me 100 per cent, even though we both knew that some people would not appreciate former chiefs speaking up on climate, and that it was likely I would have to endure personal attacks. We decided that this was a small price to pay if my colleagues and I could help get the message out about the climate emergency we clearly faced.

I hit the phones to former colleagues and floated the idea of the proposed new group, Emergency Leaders for Climate Action (ELCA), expecting that I would get quite a few knock-backs. I was wrong. Within a short period, ELCA had twenty-three members – all experienced, highly respected former emergency service leaders with expertise in disaster planning and response, whose voices could lend further credibility to the very clear but nevertheless contested evidence that increases in greenhouse gases due to human activity were causing changes to the climate that were in turn resulting in worsening fire weather. Ultimately our group grew to thirty-four members. They included former chiefs and deputy chiefs from every urban and rural fire service in Australia, former chiefs of most State Emergency services, former fire managers from forestry and national parks agencies in NSW, Victoria and Tasmania, and former directors general of Emergency Management Australia. A couple of people declined, because they felt conflicted by work they were undertaking for federal and state governments, and two others from land management agencies told me that while they shared concerns about climate, they were associated with a group pushing a strong message about the need for increased hazard reduction burning, and didn't want to mix the two messages.

What united and motivated ELCA members was feeling a moral obligation to speak out about the increasing danger Australia was facing because of escalating impacts of climate change, about our concerns that future fires and natural disasters would outstrip the capacity of even the best-resourced emergency services, and about mistruths being perpetuated by some climate-change-denying members of our federal government. We had effectively been silenced when in our former roles, often through self-censorship, as we watched with horror the political climate wars that continue to hold current and future generations of Australians hostage.

Our strength lay in the knowledge, experience and credibility of our members – that as former firefighters and emergency responders we were members of the most trusted professions – and in our simple, direct, sometimes blunt messaging, we explained how climate change was and is super-charging extreme weather events, in turn driving worsening bushfires and other natural disasters, thereby placing all Australians at increasing risk. We said that the federal government needed to start taking climate change seriously and commit to decisive action to drive down greenhouse gas emissions for the sake of future generations. Phil Koperberg put it succinctly in a radio interview in 2019, saying that we couldn't afford to keep kicking the climate-change can down the road for somebody else to deal with.

We began to actively campaign to achieve our goals, starting by warning that a disastrous fire season was likely in 2019–20, and that the federal government needed to do more to help the states and territories to prepare and respond. Under the Australian Constitution the states and territories have responsibility for emergency response; however, during a national emergency the federal government has powers, and is expected, by the public to assist. We pointed out that the government had continually declined a standing request from fire chiefs, supported by a detailed business case, to update the amount of funding originally committed back in 2003 to support the lease of large firefighting aircraft. A meagre $11M had been requested, but denied. We also pointed out that military capabilities would be required to support emergency response and community recovery, and that the processes to access this support were outdated, convoluted and slow.

In early April 2019, when we launched ELCA, we published a full-page newspaper advertisement in *The Age*, calling on the government to act on the bushfire threat and on emissions. We held a press conference of former fire chiefs in Melbourne that received wide national coverage.

I had written to Prime Minister Scott Morrison on behalf of ELCA outlining our deep concerns about climate change and Australia's ability and preparedness to deal with escalating extreme weather events, warning that we were facing a serious fire season. We requested an urgent meeting with him and followed up the request with a further letter in May 2019 after the national election, then again by email.

We heard nothing until July when the PM wrote. Curiously, the letter thanked us for congratulating him on his election win, not referring to the urgent matters we had raised. The letter said that he was too busy to meet us and in a later email his office advised that he had delegated the matter to Energy and Emissions Reduction Minister Angus Taylor, who would contact me.

It was months before any contact was made by Minister Taylor's office, coincidentally, or perhaps not, on the same day that adverse media coverage and questions in parliament arose, criticising the PM for not meeting with retired fire chiefs to hear our warnings. In fairness, it appeared that Minister Taylor may not have been aware of the matter having been referred to him. Following five months of being ignored, suddenly there seemed to be a sense of urgency in Canberra to meet with us. It was amazing what a bit of adverse media coverage could achieve with a government that seemed so committed to having a polished media image.

We tried to impress the urgency of a meeting involving the Prime Minister and the Emergency Management Minister, given our immediate concerns about the now escalating fire season. We pointed out that property was already being lost to major fires, and what we had warned of was unfolding exactly as predicted. Our request for an earlier meeting, citing the escalating bushfire crisis, seemed to fall on deaf ears, and a request to the PM's office to help us coordinate a meeting, after Minister Taylor declined to do so, never received a response.

Emergency Management Minister David Littleproud told the media that he had not been approached directly by ELCA, but would 'reach out', resulting in a meeting being arranged for 3 December. When it was confirmed that no other ministers or the Prime Minister would be in attendance, a number of ELCA members were of the opinion that we should not waste our time and money, but ultimately we decided to go ahead.

A couple of days prior to the meeting at Parliament House in Canberra I checked on timing. My understanding was that the meeting would be for an hour, but I was told firmly that, no, the meeting would not be for an hour because the minister had 'a very busy schedule', and that we had been allocated just twenty minutes, perhaps thirty if we were lucky.

There was significant cost involved in air fares and some accommodation so that retired fire chiefs from Victoria, Queensland, Tasmania and New South Wales could meet with the minister. I requested that staffers ask the minister to rethink his schedule and allow an hour for the important meeting, given that the fires were still escalating and we would be suggesting practical measures to assist beleaguered states and territories.

The meeting ultimately went for an hour. Shortly after seeing Ministers Taylor and Littleproud, and while we were still inside

Parliament House, Mr Littleproud referred to the meeting in a press conference, saying, 'They [ECLA] can take great comfort and great pride in the new breed of fire commissioners. They have planned meticulously, meticulously, for this fire season.'[23] We had no reason to be, nor had we ever been, critical of current fire chiefs. They were doing a magnificent job under the most trying conditions. Our messages on climate change and resourcing for the fires were different to and complementary to theirs. I had personally checked with AFAC to ensure that everything we said was helping, not hindering them, and AFAC was given advance copies and the opportunity to comment on many of our submissions and letters.

Former Queensland Fire and Emergency Services commissioner Lee Johnson and I were both personally attacked by the Murdoch press, erroneously claiming that we were merely 'urban firefighters' with no background in bushfire fighting – an easily refuted assertion given our decades of fighting bushfires and leading major responses, but we chose to ignore them and some other outspoken media personalities, including one who loudly complained that former fire chiefs 'who commented, outside their area of expertise, about an alleged relationship between bushfires and climate change'.[24]

On 21 November 2019, the Prime Minister was interviewed about the fires by journalist Sabra Lane on ABC's *AM* radio program:

> *Sabra Lane: 'To the fires. It's already been a very bad season. There are about three emergency alerts this morning on the Eyre peninsula in South Australia, Victoria's on code red today. The former fire chief Greg Mullins, he tried to meet with you in April to warn you that this season would be very bad and that fire seasons in the North and Southern hemispheres are overlapping making it increasingly difficult to source big water bombers. Why didn't you meet with him?'*

Scott Morrison: 'This is the advice we already had from existing fire chiefs, doing the existing job. This is why we put the additional resources into our emergency services and our aviation firefighting assets and these are things that were very well known to the government.'

ELCA lodged requests under Freedom of Information (FOI) legislation for any diary entries, attendance lists, notes or briefings about meetings with current fire chiefs or additional funding for firefighting aircraft between April and August 2019. This was the relevant period, between our first approach to the Prime Minister and when significant fires started in Queensland and New South Wales.

The FOI request regarding evidence about meetings with fire chiefs was returned to us with the following conclusion: 'I am satisfied that the Department has taken all reasonable steps to identify documents relevant to your request and that no documents relevant to your request exist.'

Similarly, the FOI request regarding additional funding for firefighting aircraft was returned to us as follows: 'I am satisfied that the Department has undertaken reasonable searches in relation to your request and that no documents were in the possession of the Department on 9 December 2019 when your FOI request was received.'

The responses to our FOI requests established that no records of the Prime Minister receiving formal briefings about the fires, or of additional funding for aircraft during the time in question, were able to be found. The Bushfire Royal Commission later established that the National Crisis Committee had only two meetings during the bushfire emergency, on 11 November 2019 and 10 January 2020.[25] On 12 December 2020, the government announced additional funding for aircraft in line with the 2018 request by AFAC.

ELCA was pushing the message that climate change was driving worsening fires – something that Canberra did *not* want to hear – and which current fire chiefs could not raise due to the vicious politicisation of climate change. AFAC had previously prepared professionally researched speaking points for fire chiefs on escalating climate threats, but the bitterness of the climate debate in political circles made it difficult for serving officers to speak up. I had been in exactly the same position back in 2009 and up until my retirement in 2017. Despite this, fire chiefs including Shane Fitzsimmons in NSW, Steve Warrington in Victoria and Greg Leach in Queensland were vocal about the unprecedented weather, fire behaviour, and some of the myths surrounding fuel loads and hazard reduction burning. Shane quite rightly called out the PM on his unilateral announcements on another day of serious fire weather, 4 January 2020.[26, 27]

Nothing concrete came from our meeting at parliament or the detailed letters with practical suggestions sent to the government before and after the meeting. I believe that Minister Littleproud understood the problematic funding issues fire services faced when trying to access additional large firefighting aircraft from overseas, and that he was ultimately able to shift the government from an intransigent position of refusing to assist the states and territories with further funding. Unfortunately, this proved to be too little too late, and made negligible difference to firefighters and impacted communities. The federal government held back two tranches of additional funding until December 2019 and January 2020, leaving little lead time to source and deploy additional aircraft from overseas. None of the additional large fixed-wing aircraft were available on the worst days from New Year's Eve 2019 through to the first week of January 2020.

ELCA continued to explain what was happening and how climate change was exacerbating bushfire risks leading to larger, more intense bushfires more likely to take lives and destroy homes.

We were dismissed as 'probably time wasters' by the Deputy Prime Minister Michael McCormack, who also said that suggestions that climate change was a factor in the fires were 'the ravings of some pure, enlightened and woke capital city greenies'. New South Wales Deputy Premier John Barilaro, also a National, said that anybody who spoke about climate change during a bushfire crisis was a 'bloody disgrace'.[28]

Given that the assumed core constituency of the Nationals are farmers and people in rural areas, it was surprising, months after the fires, when Minister Littleproud criticised the National Farmers Federation for calling on the government to adopt a target of net zero emissions by 2050, admonishing the farmers for not having a plan.[29] As I pointed out in an opinion piece in the *Sydney Morning Herald* at the time, farmers are not particularly 'radical folk', and in my experience national plans are generally expected to come from the national government.

It was a frustrating, difficult time, but a number of journalists, such as political commentator Michelle Grattan, would later reflect that ELCA had shifted the political and public dialogue about climate change and bushfires.[30] In particular, we called out the often-used government media line that it was 'not the time' to discuss climate change while fires were burning. As ELCA and bushfire survivors well know, it was *exactly* the time that communities wanted to know what was happening and why, regardless of the federal government trying to deflect debate that would draw attention to its inaction on emissions.

Nevertheless, as community knowledge and concerns shifted, so too did the Prime Minister's rhetoric, cycling through being dismissive, to acknowledging that climate change is a factor, to it being a given.[31, 32] Hopefully, the shift in rhetoric will one day be matched by changes to energy and climate policies that presently are among the most ineffective in the world. At the time of writing, in mid-2021, there are no new credible emissions or renewable energy policies, taxpayer dollars instead being invested into more gas-fired power. Methane itself is a potent greenhouse gas and its production and transport results in significant leakage. When it burns, it produces CO_2.

On behalf of ELCA and with expert input from members, I wrote detailed submissions to the Bushfires Royal Commission, a Senate Inquiry, bushfire inquiries held in New South Wales and Victoria, and we provided expert witnesses to give oral evidence. The final reports of all of these investigations highlighted what we and climate scientists had been saying – that the weather that drove the fires could not have occurred except for the influence of climate change, and that warming is being caused by humans through the burning of coal, oil and gas. After spending so many weeks and months writing, giving interviews and addressing varied groups, as well as fighting fires, I felt validated and relieved by the findings of the independent inquiries.

Ultimately, I think any report card on ELCA will show that we were able to explain to the public how climate change is stacking the odds against fire and emergency services, how this means that more people will die and more will become homeless as weather worsens into the future, and that natural ecosystems in some places are facing collapse. ELCA members find these facts to be deeply disturbing and unsettling, which is why we've refused to be compliant 'quiet Australians', a term the Prime Minister had often used, but of course in a different context.

Prior to our 3 December meeting with Ministers Littleproud and Taylor at Parliament House, the federal Opposition Leader,

Anthony Albanese, wrote to the Prime Minister seeking a bipartisan approach to the fires, suggesting a special meeting of the Council of Australian Governments, then the peak intergovernmental forum in Australia. Mr Albanese believed that there were many things that needed to be worked through by state, territory and federal governments to improve response and recovery. The PM rejected this bipartisan approach.

At an ELCA press conference in December 2019, I committed to holding a national bushfire and climate-change summit given what we termed a national leadership vacuum on bushfires and climate. The Climate Council immediately stepped up to assist, a number of other groups offering in-kind and funding support.

So as not to detract from ongoing response and recovery efforts, we determined that the summit should be held once the bushfire season had finished. Then COVID-19 hit. This left us scrambling to work out how we could hold a meaningful event.

We committed to holding a series of six virtual forums in June and July 2020. Our opening panel was facilitated by highly respected former ABC journalist Kerry O'Brien, and I was joined on the panel by former chief of the California Department of Forestry and Fire Protection, Ken Pimlott; Indigenous cultural burning expert Oliver Costello; former head of Defence Preparedness Cheryl Durrant; former CEO of AFAC Naomi Brown; Climate Councillor Professor Lesley Hughes; and Amanda McKenzie, CEO of the Climate Council.

Thousands tuned in online for the opening panel, then more than 150 experts from widespread backgrounds including doctors, farmers, fire survivors, conservationists, cultural burning experts, economists, vets, the social services sector, firefighters, insurers, scientists and defence experts participated in the series of targeted forums.

The summit produced the Australian Bushfire and Climate Plan, with 165 detailed recommendations covering issues as diverse as mental health support, telehealth, community resilience, improved response to fires, insurance, improved recovery arrangements, protection of wildlife, and securing funding to assist communities in dealing with climate change.

The report was provided to the Bushfire Royal Commission, to Prime Minister Scott Morrison, and to state, territory and federal ministers, the opposition and crossbenchers.

ELCA personally briefed Energy and Emissions Reduction Minister Angus Taylor, and later the minister for Emergency Management, David Littleproud. Unfortunately, the PM did not respond to our offer of a briefing. The ministers were quite interested in certain aspects of the plan, and Minister Littleproud requested that former Queensland Fire and Emergency Services commissioner Lee Johnson and I speak to him again once the Bushfire Royal Commission handed down its report.

I wrote to Minister Littleproud ahead of this planned meeting to ask a number of questions about the federal government's detailed response to the Royal Commission report. While he and the PM had told the media that they broadly accepted its eighty recommendations, the government's written response stated that it 'supported' thirty-two recommendations, 'supported in principle' twenty-five and 'noted' twenty-three. I asked what this actually meant. Media reports of previous comments I had made about the Royal Commission and climate change seemed to incense Minister Littleproud, who told the media in October 2020 that I was being 'disrespectful to our current fire commissioners in effectively claiming that they did not provide that advice'.[33]

One of the Royal Commission recommendations merely 'noted' by the government was to develop a 'modest' locally based large firefighting aircraft capability, reflecting issues detailed by fire chiefs in their 2018 business case, that non-availability of aircraft was a result of overlapping fire seasons, that there was insufficient funding to have enough aircraft based in Australia, given escalating fire risks, and that the federal government had not updated its contribution to the national firefighting fleet since 2003. ELCA had assisted by reinforcing those concerns.

Minister Littleproud's initial response to our detailed letters and advice was a disappointing formulaic response. Then on 9 December 2020 we received another more detailed letter – dated 8 December – from the minister. It talked in the past-tense about the telephone meeting scheduled for 10 December which had yet to take place, referring to what had been discussed. Then, at short notice, Mr Littleproud pulled out of the 10 December phone conference he had originally requested, promising that senior staff would be on the phone instead. On 10 December at the appointed time former commissioner Johnson and I listened to on-hold music for fifteen minutes while we arranged for somebody else to call the minister's office to remind them of the 30-minute phone hook-up. Eventually we had a very fast, unsatisfactory chat to a staffer.

The 8 December letter could have been very embarrassing for the minister and government had it been released to the media. Instead, I chose to contact a senior public servant, who went to significant effort to arrange a substantive response to the issues we had originally raised.

It is unfortunate that the government failed to listen when bushfire experts tried to warn it of the approaching disaster and that it failed for months to take practical measures that may have lessened the impacts of the worst bushfires ever experienced. Hindsight is an inexact science, but no amount of reflection and deflection changes the fact that the government was warned, and history will record that it chose not to act until far too late.

On the issue of emissions, which drive warming and extreme weather, the policy vacuum continues. As I stated in several interviews during the fires, the best analogy is a pot boiling over on a stove. The only way to stop it boiling over is to turn down the heat. Our escalating bushfires are the same – they have reached proportions that make them impossible to control on the worst fire weather days. In the previous century we had become adept at fighting bushfires in the high 30s and low 40s, then our worst fire days. Fires burning on days approaching 50°C are beyond anything that humans have previously experienced, and beyond the capacity of current firefighting arrangements and technologies. Therefore it is imperative that we dial down the heat, by tackling emissions, or the pot will keep boiling over. The Bushfires Royal Commission report said that decades of warming is locked in because of what has already been emitted, and conditions will worsen as a result regardless of what we do. But, it warned, what happens in the future is entirely reliant on what worldwide governments, including ours, do about emissions reduction *now*. If the current inaction continues, fire seasons like Black Summer are likely to be common by 2040, and the new normal by 2060.[34] The worst seasons will be worse than anything we can currently imagine, which is frightening to contemplate.

7

BLACK SUMMER

ULTIMATELY, ELCA HAD TO face the worst possible outcome for us. Simply put, we were right. The fire season we faced proved to be even worse than we feared. We had warned from April 2019 that a catastrophic fire season was approaching, but ultimately all of us were shocked at the sheer size, destructiveness, scale of loss and length of the worst fire season ever experienced in Australia. History will remember it as Black Summer.

Black Summer 2019–20 wasn't really a summer, but a fire season than lasted more than half a year. Fires started burning in July, winter, then burned into spring and summer until flooding rains arrived in late February and early March 2020. The fire season began nearly three months earlier than what had been normal in the nineteenth and twentieth centuries, and reflected the trend of lengthening fire seasons since the start of the twenty-first century.

The fire season barely took a breath in 2019. The RFS reported that more than 1,000 fires were recorded in each of the months of June, July and August even though the official bushfire danger period doesn't start until 1 October. Many local councils in New South Wales brought forward the statutory bushfire danger period to 1 August, with the rest of the state following suit on 1 September.

The Lindfield Park Road fire started near Port Macquarie on the Mid North Coast on 18 July. It would prove to be stubborn, burning in peat below the ground and regularly popping up, setting grass and scrub alight when conditions were hot, dry and windy. Underground fires are becoming increasingly common as swamps and bogs dry out. They are notoriously difficult to extinguish, with flooding the only proven remedy. This fire would continue burning until heavy rains in February.

August saw large fires breaking out in warm, dry, windy conditions at multiple locations in the north of the state. Fires once again broke out in the Northern Tablelands near Tenterfield where they had burned earlier in the year, to the south near Armidale, and on the North Coast near Yamba.

What fire authorities had feared was now coming to pass. As with the previous year, we were witnessing yet another early start to the fire season in line with the long-term warming and drying trend, plus the failure of winter rains for the third year in a row. New South Wales was in the midst of the worst drought in living memory, even worse than the longer Millennium Drought at the turn of the century. Periodic heatwaves and an increase in average temperatures year-round meant bushfire fuels were even drier than would have been expected in previous droughts, with higher temperatures increasing rates of evaporation, and low humidity preventing moisture recovery in vegetation.

Even rainforests were burning intensely in places where there had never been intense fires before, such as the ancient Gondwana rainforests of the Hyland Nature Reserve near Dorrigo, and subtropical rainforest in southern and central Queensland. It was still winter, but we were seeing fire behaviour similar to what could be expected in late spring.

Late in the evening of 18 August a number of emergency calls were received reporting a bushfire in Ku-ring-gai Chase National Park on the western shores of Pittwater opposite Palm Beach. A couple of RFS tankers responded; their crews could see an orange glow, but could not safely access the fire. Driving on fire trails, they estimated that they were still about a kilometre from the fire. It was decided to wait until daylight to deploy firefighters and aircraft, because the terrain was steep and dangerous.

The next morning, I headed off to the fire in our brigade tanker, together with about five other RFS tankers and a number of National Parks and Wildlife Service crews. A National Parks helicopter was also assigned to the fire, and two RFS fire boats worked from the shores on Pittwater.

The officer in charge of one of the other crews was Bob Conroy, whom I had known for years, having worked together on many of the same fires. We often consulted each other on how fire seasons were shaping up, and I trusted his take on issues, given his experience as a firefighter and his environmental knowledge. Since retiring from the NSW National Parks and Wildlife Service and becoming a volunteer, he had deployed to bushfires in Tasmania, in the USA and in Queensland. A very fit, intelligent and quietly spoken person, Bob is revered in his RFS brigade and by his former colleagues for his depth of bushfire knowledge.

The fire at Longnose Point was in a remote spot and investigation suggested it was the result of an escaped campfire. Somebody had come ashore the previous afternoon, climbed up steep cliffs to sandstone caves, and failed to properly extinguish their illegal campfire before leaving. In the tinder-dry conditions, only a spark was needed to ignite the surrounding bush. The fire had burned uphill to the top of the ridge and would have been highly visible to the many residents who lived on the eastern side of Pittwater.

When the crews had tried to gain access the previous evening, a very dry but thankfully light westerly wind was blowing, which meant that the fire was burning against the wind and being slowed down, rather than accelerated, when it reached the ridge top. The risk versus reward of sending firefighters into thick bush to find their way to the fire, navigating sandstone cliffs and steep hills in the dark, was not worth it.

I remember driving in convoy along the fire trails for a couple of kilometres, watching the smoke as we approached. I was fairly comfortable with what I saw. Light brown and white smoke, with not too much heat under it. Despite this, I got the crew involved by asking them to point out potential refuge areas and turn-around points.

Another lesson from my dad is never, ever assume that everything will go perfectly to plan: be the eternal pessimist, planning for the worst while hoping for the best. I know that this message saved my life several times over the years, and I had always tried to instil the same level of calm preparation and planning in other firefighters I was responsible for.

We arrived at the end of the fire trail, a perch with magnificent views to the east over Pittwater and Palm Beach from the tops of jagged sandstone cliffs, about 150 metres from the closest edge of the fire. It was relatively simple to deal with. In normal circumstances we would have employed dry firefighting, which is when you rake trails around the fire edge to starve it of fuel. However, we knew that because of the extreme dryness, logs and trees would continue to smoulder and the slightest puff of wind could send sparks across control lines to start new fires over coming days and even weeks.

To mitigate that risk, the plan was to run hundreds of metres of hose through the bush and mount a direct attack on the flames, which were burning against the wind and were only 1–2 metres high, then scrape a trail along the burnt edges so the fire couldn't rekindle and take off again. We went in, and our small army of park rangers and volunteer firefighters started to come to grips with the fire.

As we got to the brow of a hill after extinguishing about 200 metres of fire front, I looked down a very steep slope punctuated by sandstone cliffs. We had to rig ropes at some of the cliff lines so that we could clamber up and down. The terrain was extremely steep and dangerous, and I reflected on the good decision of the previous night not to send people in.

While we mounted a direct attack on the western flank of the fire, the fire boats were running hoses uphill from the waterline below us at two locations. The plan was to pinch out the fire fronts by continuing to add 30-metre lengths of hose until we met up with the boat crews working uphill towards us. We started to experience water supply problems due to friction loss in the long hose lines, then rigged up portable pumps to boost the pressure. Later, as we worked downhill, the pressure became too high and we had to turn off one of the pumps. Firefighters need to have a working knowledge of many things, including basic hydraulics.

Up above, the National Parks helicopter with a water bucket was doing pinpoint drops onto areas of heavy fuel whenever the fire intensified. This was crucial because some of the downhill sections of the uneven fire edge meant that sometimes we were working above active fire fronts, which was both unpleasant because of the smoke, but also potentially dangerous if the fire flared up directly below us. A couple of times just a breath of wind sent flames that previously had been around 1–2 metres in height up to 4 metres, sometimes flaring into the treetops for a moment.

We weren't particularly worried. The weather forecast was benign – warm and dry, but not too much wind – and we had a good plan to round the fire up. We employed the technique of 'one foot in the black', which meant that we stayed on the fire edge, and if there was a flare-up or conditions deteriorated, we could simply run onto burnt ground where the fire could not reach us because there was no more fuel left to burn. Everyone had to be on their toes because occasionally large tree branches and the odd tree would burn through and crash to the ground. We were constantly looking up, assessing the trees around us.

As crews worked along the fire edge, knocking down flames with the hoses, we all took turns working with McLeod tools (rake-hoes), scraping the extinguished fire edge down to bare earth to create a fire break about 2 metres wide. Later in the day we spent a lot of time wetting down everything within at least 10 metres of the fire edge – mopping up – and paid particular attention to burning banksia cones, tree branches and hollow trees – anything up high that might burst into flames if the wind came up, drop sparks across the fire line, and send us back to square one.

Mopping up is dirty, wet, arduous work, and often very unpopular. I remember my dad many times over the years, when people complained, saying, 'Well, by all means let's all go home. But I'll see you back here in a few hours when it jumps the line again. Maybe it's better that we put the effort in now so we can stay in bed tonight?' I never forgot that advice. Saving a little bit of time and effort now would probably mean far more work and damage further down the track.

Bob and I scouted ahead a couple of times. He had been appointed as division commander of the western edge of the fire, and gave calm, clear instructions to the sector leaders. We talked about what was happening up north with the fires that had started to burn in July, and how dire it was starting to look due to the relentless westerly winds and continuing drought.

I'll never forget Bob saying, 'I think this is the year, Greg – it's going to be the big one. The Keetch-Byram Drought Index is already over 100 in most of the state and look at the weather we're getting.'

There are not that many people in the fire business with Bob's level of experience and scientific understanding of the environment and fire risk. Our careers had been quite similar in many ways. Like me, he had been overseas on a Churchill Fellowship studying bushfires, we both had master's degrees in management, we had lived and breathed bushfires over our long careers, and we were both back on the front line as volunteers following long careers and formal retirement. We both shared deep concerns about what climate change was doing to long-term trends in bushfire risk and both felt that we were duty-bound to do everything in our power to tell people about it.

Bob gave voice to what he was seeing happen around us that day, and I agreed with him 100 per cent. The bushfire season of 2019–2020 indeed looked like it was going to be *the* year, and as members of ELCA we had been doing our best since April to draw attention to the extreme level of risk for the looming fire season, but also to what the future might hold if urgent action on emissions continued to be deferred. Neither of us realised at that stage though just how bad the season would turn out to be. History was no longer an indicator of the future, and what unfolded was shocking. We later reflected that we wished we had been wrong.

Major fires continued to burn in the north of New South Wales through September and October. I remember speaking to Erris about what was happening, and warning that it was inevitable that I would be called away to fight fires, sometimes for days at a time.

As always, she was supportive, but also apprehensive, because she recognised that the weather and fire conditions in the first week of September were already proving to be worse than most summer fire seasons. She impressed on me that I needed to realise I was no longer as young as I used to be, and that while I had lots of experience, this season warranted me being extra careful. I missed not being able to talk to Dad about the conditions and to get his thoughts, but I also knew that if he was still with us he would have echoed Erris's 'safety first' message.

On 5 and 6 September many weather and fire danger records were smashed, or, as the official reports stated, 'exceeded by large margins'. Locations in northern New South Wales and southern Queensland experienced temperatures up to 10°C above average, accompanied by the highest September fire danger indices ever experienced since records began in 1950. On 6 September, strong west to north-westerly winds and high temperatures resulted in widespread Extreme Fire Danger with pockets of Catastrophic Fire Danger in Warrego and Maranoa, the Darling Downs and Granite Belt districts in Queensland, and north-eastern districts of New South Wales.[35]

The Bureau of Meteorology later reported that conditions preceding the blast of unprecedented early fire weather were the driest on record in some of the affected areas, with the New South Wales Northern Tablelands experiencing 77 per cent less rainfall than the long-term average, and the lowest June to August rainfall on record in the Southern Downs (Queensland) and Northern Tablelands.

The extreme dryness together with strong winds and high temperatures resulted in critical fire situations in many areas. Lightning had started a fire near the township of Drake on 4 September, the Long Gully Road fire, and on 5 September it raged out of control and destroyed properties. At the same time, further south, the

Bees Nest fire also ran out of control and damaged and destroyed property. This fire consumed ancient and rare Gondwana rainforest that had been untouched by intense fire for millions of years.

In Queensland, many fires burned out of control from 2–13 September, with major fires impacting the Scenic Rim area (eleven homes and five commercial buildings destroyed), the Southern Downs / Stanthorpe region, and Peregian Beach (one home destroyed). One of the casualties of the fires was the iconic Binna Burra Lodge and cabins, perched on a mountaintop in pristine rainforest, formerly thought to be too wet to burn. The resort was destroyed.

The 2019–20 Queensland bushfires would prove to be the most destructive on record for that state, with more than 7.5 million hectares burned by around 3,000 fires, forty-nine homes destroyed, 100 damaged, seven public assets such as schools damaged, and many national parks impacted.[36]

The fires in northern NSW, despite being so early in the season, stubbornly resisted all efforts at control. They grew bigger as periodic days of high winds and high temperatures made them rage out of control. In the intervening days of milder conditions, firefighters did the best they could, but the sheer number of fires, ever-growing fire perimeters and inaccessibility of wilderness areas made their jobs all but impossible.

By the beginning of October, large fires were well established in northern New South Wales and firefighting authorities were stretched thin. Periodic episodes of Very High Fire Danger and above saw fires expand significantly, with property losses mounting. On 8 October two people lost their lives in the ongoing Long Gully Road fire. That day the small town of Rappville was heavily impacted with the loss of forty homes and the biggest local employer, the sawmill.

A couple of weeks after the fire burned through Rappville, I worked on the Long Gully Road fire for a few days. At one stage we had to deal with a fire that had rekindled well within the burnt area and was burning through a carpet of dead leaves that had fallen to the ground after the main fire two weeks before. None of us had ever seen this before. Weeks after this fire was thought to be out, it would again rekindle in the extreme dryness. This turn of events underlined just how difficult the task of firefighters was becoming under conditions of long-term rainfall reductions, extreme drought, extreme temperatures, high winds and low humidity. A lot of tried and true firefighting strategies and tactics were not working as they had in the past. As the fires multiplied and grew, firefighting resources were stretched thinner and thinner across the state. Each night back at the base camp I would call Erris to let her know that I was okay, assuring her that I remembered what she'd said about staying safe. I know that many firefighters on occasion decide that they need to be selective about how much they tell loved ones about what they have faced. This was certainly the case for me on the worst days during Black Summer.

In late October 2019, more dry lightning storms arrived, starting new fires in remote areas. Because of the stretched fire-fighting resources, including availability of the limited number of aircraft, it was not possible to contain all of the new fires and in the following days and weeks some grew significantly whenever there were strong winds and high temperatures.

On 26 October, lightning started the Gospers Mountain fire in a remote wilderness area to the north-west of Sydney. The fire would eventually grow to become the largest forest fire in Australian history, burning 513,000 hectares and destroying about 100 homes and many other buildings such as farm sheds. It would merge with several other large fires, including the Kerry Ridge,

Three Mile, Grose Valley, and Little L Complex fires, in total consuming more than 1 million hectares.[37]

In December, former prime minister Tony Abbott and I worked together as members of a strike team on the Three Mile fire northwest of Sydney on a day of Severe Fire Danger. At one stage as the wind increased in velocity, the fire jumped Putty Road from west to east. Tony and I positioned our crews on opposite ends of the large spot fire and kept in touch by radio. We tried in vain to work around the perimeter of the fire and stop it spreading east, as strengthening winds and a steep slope saw the fire escalate rapidly. Tony and his team had the worst of it – intense flames outflanked them, burning through their hoses.

A very capable and experienced firefighter, Tony immediately evacuated his crew onto burnt ground as we tried to reach them. We too were forced to watch from the safety of burnt ground as the fire roared into the treetops and raced up a steep hill, out of reach of our hoses. Later we would set up to protect isolated rural properties in the fire's path as it burned to the east, eventually joining up with the Gospers Mountain mega-fire.

I have high regard for Tony and his community spirit and have been to several large fires with him. I have sometimes been asked how I reconcile our diametrically opposed views on climate change. The answer is that we choose not to discuss it because we know neither of us will convince the other to change sides in the climate debate. One of the things that I despair about in Australian politics is the deep divide over climate and the resulting inability for either of the major parties to progress meaningful action on emissions.

I wish that the culture in Canberra would focus more on agreement than disagreement and be generally more respectful, seeking out the bipartisanship that has seen us through wars and the recent

COVID-19 pandemic, an approach that in other countries is seeing real action on emissions.

As our fires continued to grow, blackening huge areas, killing native animals and destroying more property, fires were still burning on the other side of the world. This underlined the increasing overlap of fire seasons between the northern and southern hemispheres, something that is now having an impact on the availability of specialised people and equipment that have been routinely shared over the years, particularly large firefighting aircraft.

In late October, major fires broke out in California: in Orange County near Los Angeles and Sonoma County north-east of San Francisco. The communications director for the Climate Council, Lisa Upton, a seasoned journalist who had covered humanitarian crises in some of the world's most dangerous hotspots, thought that this could be an ideal opportunity to draw attention to how climate change is impacting worldwide, and how international firefighting arrangements are being affected. With the support of the Climate Council, Lisa and I quickly arranged to fly to the USA.

Before leaving, I spoke to two friends, James Featherstone, former Los Angeles fire chief and general manager of Emergency Management, whom I'd known since 1995, and Ken Pimlott, recently retired fire chief and head of CAL FIRE, the Californian Department of Forestry and Fire Protection. James and Ken opened doors and arranged for me to speak to key people involved in managing two large fires.

On arrival in San Francisco, I spoke to the deputy fire chief in Orange County and learned, thankfully, that fire and weather conditions had eased and there was no point heading there. CAL FIRE, however, still had a lot of work to do to contain the massive Kincade fire, so we arranged to head there the next day together with a locally hired camera operator. The ABC contacted us and asked if Zoe Daniel, the US bureau chief based in Washington DC, could accompany us to fires.

Over the next few days we had some fascinating discussions with senior fire officers, frontline firefighters and a renowned climate scientist from Stanford University, Professor Chris Field. Zoe and her crew followed us into the mountainous fire areas and burnt-out vineyards to put together a piece for the ABC's flagship current affairs TV show, *7:30*.

Local firefighters told me how, in that part of northern California, fire seasons had historically been quite short and big fires rare. In the last couple of decades, though, they said they had been smashed, with drought and with large, destructive fires almost annually, ever more numerous property losses, and, in 2018, the loss of nearly 100 lives as the Camp fire obliterated the city of Paradise.

I heard how their fire seasons were now months longer than twenty or thirty years ago, and it was not uncommon to have simultaneous major fires in both southern and northern California, something that was previously rare. During the Kincade fire, unprecedented wind speeds of 129 kilometres per hour pushed the fire towards the coast and hundreds of thousands of people in the path of the fire were evacuated.

Ken Pimlott had previously explained to me that reduced snowpack due to climbing temperatures had changed the water cycle in many parts of the state. Less snow meant less runoff and lower river flows, with a consequential impact on ground water and soil dryness. Many people expressed the view that California now seemed to be in an almost constant state of drought, exacerbated by extreme temperatures that increased evaporation rates. Professor Field explained to me that due to the warming climate, fire was a tool being used by Mother Nature in some places to change forests into scrubland, scrubland into grassland, and grass-land into desert.

On return to Australia I went back to fighting fires. The *7:30* episode aired nationally on 6 November, detailing parallels between changed fire conditions driven by climate change in both the USA and Australia, highlighting that this was now an escalating world-wide problem.

Standing in the ruins of a burnt home in Sonoma, I said, 'This is the worst build-up to a fire season I've seen since 1994, when New South Wales was devastated.' Two days later, on 8 November, extreme fire weather saw major fire runs and property loss in northern New South Wales and southern Queensland, validating the fears I had expressed to the ABC.

On the east coast of Australia, fire weather moves progressively south during most fire seasons, from Queensland to northern New South Wales, to the Mid North Coast, Sydney and environs, then the Australian Capital Territory, south coast and on to the southern states (Victoria, South Australia and Tasmania) from January, with February usually the worst month for fires there.

By early November 2019, the New South Wales fires had become well-established in the Clarence and Richmond Valleys to the north, and fires were burning from the Queensland border all the way to the Blue Mountains west of Sydney where the massive Gospers Mountain fire continued to grow.

Extreme fire weather was forecast for 8 November. I recall thinking on that day that the fire season had transformed from something we had experience of in the past to something frighten-ing and new. Even I was shocked at the eventual toll and the level of destruction in areas where, to my knowledge, we had never before experienced significant property losses from fires. The toll on the natural world was immense, with fires burning through landscapes previously almost immune to fire for millions of years. This was the time that the word 'unprecedented', a term often misused to exaggerate a situation, started to pop up in the media. This time there was no exaggeration.

Starting on November 8, the Mid North Coast area of New South Wales suffered significant property losses with more than 150 homes and almost 400 other buildings destroyed. Many more were seriously damaged. Dozens of homes were destroyed to the north-west near Glen Innes; a school was lost and about sixty-two homes were destroyed in the town of South Arm on the north coast near Macksville.[38] Only once before in the state's history, in 2013, had large numbers of properties and lives been lost earlier than late November, showing that serious fire weather prior to this had been rare through the nineteenth and twentieth centuries. By 8 November the number of properties lost to fire in NSW had already exceeded the previous worst bushfire property losses in New South Wales fire history, surpassing the worst previous losses in October 2013 and January 1994. Truly unprecedented, and with months of fire weather to come.

One of the people who lost her home on 8 November was Fiona Lee. She shared her story with me in the aftermath of her terrible loss.

Fiona, her partner Aaron and their daughter had moved back east after spending time in a remote part of Western Australia. Aaron's parents owned a large country property at Warrawillah about 30 minutes' drive from the regional centre of Taree on the New South Wales Mid North Coast. They were relishing being back close to family and were building their dream of living sustainably off the grid, surrounded by nature.

They established vegetable gardens, rainwater tanks, solar panels and worked hard to be self-sufficient and resilient. Their only external communication was via satellite links and through limited mobile phone coverage at the top of a nearby hill. Alternatively, they could drive towards the town of Wingham to the reception spot, a bend in the road about 20 minutes from their home.

Fiona was well versed on the impacts of climate change and had been active with Extinction Rebellion, protesting against the government's failure to take meaningful action on emissions reduction. She had followed the news about escalating fires and knew that it was not normal to have such large, destructive fires so early in the year, nor the hot, dry, windy conditions driving the fires. She knew from her efforts to establish new gardens that the ground was incredibly dry. Their rainwater tanks had not been topped up in months.

In early October, lightning started a fire in a remote mountainous area to the west of their home, the Rumba Dump fire. Over the following weeks they watched apprehensively as the smoke thickened and went dark brown on hot, windy days as the fire steadily grew. Often the area was enveloped in smoke and breathing was difficult, as was seeing anything through the constant smoke haze.

The fire was a long way away, but Fiona and Aaron knew that with the right weather conditions they could be threatened at some stage and receive very little warning. They spoke about what they would do if fire threatened, given that their water tanks were low and the surrounding bush was tinder dry. They realised that if a fire arrived they would need to act quickly if they decided to leave, as they could easily be cut off and trapped by a fast-moving fire on a bad fire weather day. Staying to defend with no water was out of the question.

On 7 November they could see the smoke plume, closer than ever, but because they had little in the way of contact with the outside world they weren't sure what they should do. They had been using the internet via the satellite link and accessing the NSW RFS Fires Near Me app to maintain situational awareness. There were no warnings that they thought related to them that day.

The following morning Fiona and Aaron woke to the strong smell of smoke. The day was already hot, with winds gusting from the west. They saw that the day's fire danger was forecast to be Severe, but having already been through a couple of days of Extreme Fire Danger in October, this in itself wasn't a trigger for them to leave. Through the morning it became hotter and hotter, and the winds picked up. The sky went dark and smoke covered the sun. Burnt leaves and ash were falling all around them and they could see towering columns of smoke reaching high up into the atmosphere. The smoke was getting thicker, but it was hard to work out just how far away the fire front was. They listened and looked out for any aircraft or fire trucks, but heard and saw nothing, hoping that this meant they would be spared, and the fire wouldn't come their way.

They were packing for a camping trip, with a 5-metre sea kayak strapped to the top of the car. As it got windier and the smoke got closer, Fiona started to get scared, urging Aaron a couple of times through the morning, 'Let's go! Let's go!' as they continued to pack. At about 1 pm, with black smoke looming over the top of them, they'd had enough. It was time to get out of there. They called past Aaron's parents' house and told them, 'We're going,' but his mum and dad said they were staying to defend their home.

About 10 kilometres down the road, towards the small town of Wingham at the reception spot, they came across a crowd of locals all calling loved ones to check on their welfare. As reception cut in, their mobile phones started to vibrate with emergency warnings saying that it was too late to leave, and to take shelter. Hopefully, Aaron's parents would receive the warning via their landline. Burnt leaves and ash were falling all around them and they could see towering columns of smoke reaching high up into the sky.

They continued on into Wingham. Aaron couldn't reach his parents so went back to the valley to find them, desperate to know they were safe. He negotiated his way through police roadblocks, around fire trucks and eventually made it to his parents' house. It was still standing, and they were okay. He realised though that his own house was probably on fire.

In Wingham, Fiona and others tuned in to the fire services' radio channels, listening for news about people's homes, hearing frantic messages from firefighters trying, unsuccessfully in some cases, to heroically save homes from the onslaught of flames. At about midnight Aaron finally called. Fiona said that when she saw Aaron's number come up on her phone, her stomach lurched. She answered, and Aaron told her that Mum and Dad were safe. He had climbed up the burnt-out hill near their house where he knew there would be some phone reception. Too frightened to ask, she waited for what he'd called to tell her. She told me that she will never forget his next words, 'It's. All. Gone.' Words that she'd heard from others that night and would hear again in following days. Everything they'd worked for. Gone. In the blink of an eye.

She told me that she had an odd feeling when he told her: relief. She knew, somehow, that the fire was going to take everything in its path, and Aaron had confirmed it. But they'd made it out with the most precious thing of all – their lives. And, fortuitously, she thought, with their camping gear! That would come in handy now that they were . . . homeless, a word that Fiona struggled to come to terms with.

Aaron returned and the fire refugees from surrounding areas continued to monitor radios, Fires Near Me, and any news they could get their hands on. Even at night it was still hot and windy. To their west they could see the orange glow from the Rumba Dump fire as it continued to burn through the mountains, consuming everything in its path. There were still emergency warnings for the nearby Hillville fire. And to their east were the Crowdy Head fire and fires around Forster, also subject to emergency warnings. Fiona and Aaron wondered if they were safe in Wingham or perhaps should try to get to the larger town of Taree, or instead head south to the city of Newcastle. Nobody could tell them. An exhausting, sleepless night in their car followed, as they contemplated their loss and what life would be like from now on.

Ultimately, with their daughter's asthma made worse by the constant smoke, they decided to get out of there and head to Newcastle where Fiona's mum had a place. 'I just didn't have the emotional capacity to stay and help – I was totally drained,' Fiona said.

There was a lot of media coverage of the fires in this part of the state. Fiona was deeply offended when Deputy Prime Minister Michael McCormack made disparaging comments about 'woke inner city greenies' and the fires, saying that people shouldn't talk about climate change when the priority was responding to immediate needs like providing shelter. NSW Premier Gladys Berejiklian weighed in and said that it 'wasn't the time' to discuss climate change. Fiona told me that when she heard ex-fire chiefs respond with, 'If not now, then when?' it encouraged her to think about what she could do to get the climate change message out.

Aaron came up with the idea of dumping some of the ash from their destroyed home outside Parliament House in Sydney to raise awareness about what was going on in the bush. Afterwards, Fiona said, 'That felt good. The politicians weren't speaking for me, and they needed a wake-up call.' Prior to this they had returned to the site of their former home with a TV crew, believing that climate change would be central to the story. The footage where climate was mentioned, however, ended up on the cutting room floor, which was upsetting. 'That was the only reason we travelled back there and reopened the wounds – to warn about climate change. But no, they didn't run any of that.'

Fiona went on to make and exhibit some artwork that told her fire and climate change story. I spoke to her again in March 2021, and asked her what plans the Lee family had for the future. She said that while things were going well – they were renting in Newcastle, her daughter was happy and well, and Aaron was studying at university – ultimately, they really wanted to get back to the bush. They planned to buy some land, but remain deeply cautious after their experience. With each property they consider, they ask themselves, how safe will it be? Can we get away with building anything other than a protective bunker? How can we make sure there's enough water? They both know that climate impacts are escalating and that projections for future fires are increasingly dire.

As she finished sharing her story, Fiona reflected, 'People are now being forced to run from the bush to cities for safety. I always thought it would be the other way around – that city living would become unsustainable, and people would increasingly leave cities for the bush, for better lifestyles, healthy food, and nature. Climate change has turned all of that right around. It felt like our security had been taken away from us. I felt that as a mother I could no longer protect my daughter. It was just a horrible feeling that I still feel to an extent.'

This was just one person's story of their life being turned upside down by bushfires supercharged by climate change, and of course there are thousands more. Fiona's story impressed on me that we are becoming utterly powerless against the forces of nature unleashed by climate change. I wondered, with some despair, when, if ever, our elected officials would grasp what we are facing and decide to do something credible about it.

8

ON THE FRONT LINE OF AUSTRALIA'S FIRST GIGA-FIRE

BLACK SUMMER ROLLED ON. On 15 December, the captain of Terrey Hills rural fire brigade, Peter Duff, was having a welcome break from fighting fires and took his two young sons out on his boat. Even though there had been no significant fires locally, Pete had been organising crews for the brigade's two fire trucks for months as well as taking charge of crews and attending out-of-area fires himself. Aged thirty-seven, Pete had been a volunteer firefighter since 2000, following his dad, Kevin, who had also been captain, into the brigade. A truck driver, Pete has a reputation as a very competent straight-shooter, and had taken on the position of captain in early 2019.

At about 1:30 pm as he was washing down the boat, his pager went off. Pete knew that both Terrey Hills trucks were already away at fires and was a bit confused. He called the fire control centre and they immediately asked him to take charge of a truck from another brigade that did not have a full crew available. There had been an urgent call for help from another district.

On arriving at the fire station a short time later he was surprised to find that they had been asked to respond urgently to Mount Tomah, located nearly 100 kilometres to the west in the Blue Mountains.

It was unusual to be asked to respond with lights and sirens over such a long distance, but Pete knew that this meant there were significant problems up there.

As the crew climbed the foothills of the mountains, the conditions were quite still – not a lot of wind – but hot and smoky. Because of the smoke haze covering Sydney, they couldn't see a distinct smoke column up ahead. They received new instructions to go to a picnic area near the Botanical Gardens at Mount Tomah, so they drove along the Bells Line of Road, noticing a large number of fire trucks heading in the same direction. Overhead, constant sorties of large and very large air tankers flew over at low altitude, obviously operating out of the nearby air force base at Richmond. Pete later told me, 'I thought to myself – they're not up there for fun. Something bad is happening.'

As they headed higher into the mountains the smoke haze cleared and ahead the volunteers saw a dark smoke column reaching high into the sky. On arrival at Mount Tomah, conditions were fairly still and for a couple of hours all they could do was wait. Pete knew though that hot, dry westerly winds would kick in at about 3 pm – as they had done for weeks – and the fires would start to run. He recalled that, suddenly, the wind came out of nowhere.

The sector commander pulled them out of their location, saying that it was too exposed and dangerous. The black smoke told them that the fire was now climbing above the treetops and moving quickly. They relocated east to nearby Berambing, setting up to protect homes on large rural blocks. The wind strengthened even more and everybody's phones pinged with an Emergency Warning saying that a fire-generated storm had developed and it was too late to leave. The warning advised people to seek immediate shelter.

Pete did a stocktake: they had hoses out, there were cleared areas around the house, and he felt they had a couple of refuges they could run to if things got dicey. He nevertheless admitted to me that he thought to himself: *This is it. I'm about to see a real fire.* This from a twenty-year veteran whose father, Kevin, had also been a brigade captain. Like me, Pete had grown up watching his dad fight fires and then followed him into the local volunteer brigade. He'd seen many fires but sensed that today was going to be different.

'Suddenly things just EXPLODED! A howling wind hit us, huge embers floating down out of the sky. And the SOUND! It was like a freight train heading straight at us,' he said.

In footage that went viral around the world, Pete filmed with his mobile phone as fire surged up out of the valley. He sent an urgent radio message to the division commander: 'Flame height is at least 60 metres, repeat 60 metres.' I asked him how he stayed so calm when he sent the message. He replied, 'Mate, I might have sounded calm, but I didn't feel it!'

The firefighters set to work as flames towered above them and the lawn around the house caught fire. They heard a constant crashing sound as trees burned through and fell to the ground in the fire-generated winds. Pete knew that there were farm buildings down near the bush, but there was no way to reach them and he wasn't going to put his firefighters' lives in danger in futile efforts to put them out. He realised that by now they would all be on fire, something he was later able to confirm. But they saved the house, which felt good when they heard that others had not survived.

The fire jumped over them and burned into the inaccessible Grose Valley. A week later, Pete and I would square off against that blaze as it threatened the town of Blackheath. He told me that he found the fire behaviour extraordinary. Just six years earlier the State Mine fire had burned through the same area. Normal expectations would be that the area would remain fire-free for a decade or more because of reduced fuel loads. But here it was, burning so intensely that it generated a firestorm, just six years after another very intense fire. The difference of course was the weather: hotter, drier and windier than ever before.

It goes without saying that the partners of firefighters are special people. I know this from firsthand experience. I saw how Mum supported Dad. Erris has always been incredibly supportive of me. Though I have to admit, when fighting fires I often don't think much about the impact on the family.

After Pete's footage of the fires went viral, Pete's wife Jen wrote something on Facebook, which itself found a wide audience:

> *These are the scenes that are terrifying us all right now.*
>
> *Especially terrifying to myself and my two little boys.*
>
> *As we waved goodbye to Daddy, Peter Duff, at lunchtime today, just as we have numerous times already over the last few months, I think to myself, 'please, please, please come home to us.'*
>
> *Then I see his video.*
>
> *I feel most people are extraordinarily supportive of what our incredible volunteers do, but these scenes show the extremely real danger these precious people put themselves in to save others and their homes.*
>
> *Please nurture these selfless people and remember that they mean the world to the ones they leave at home. We reluctantly, but graciously let them go. All while terrified they may pay the ultimate price.*
>
> *Love you all and hope with everything I've got that the conditions allow your safe return.*

*

The fires continued their march to the south, with more and more of the forested eastern parts of New South Wales succumbing to flames. Western parts of New South Wales were not a consideration, because the ongoing drought had parched the semi-arid landscape and there was no vegetation left to burn.

The Gospers Mountain fire grew larger and moved closer to the Hawkesbury River, the gateway to Sydney. There were significant fears that on a day of serious fire weather the fire could jump the river and burn through northern parts of Sydney, impacting thousands of homes in The Hills, Hornsby, and Northern Beaches districts. It also had the potential to spread east and impact on Central Coast communities, where thousands of homes were exposed to bushfire risk.

To the south of Sydney, a lightning strike started a fire in the wilderness of the Warragamba Dam catchment. The resulting Green Wattle Creek fire burned in remote areas with limited access. Previous fires in the area had proven difficult to contain and extinguish.

Lightning also sparked fires during December on the Great Dividing Range west of towns on the south coast. These remote fires continued to grow as fire weather conditions deteriorated. Those who remembered the 1968 fires which started in a similar way and in similar locations grew increasingly concerned and I remembered the dire predictions of Dr Simon Heemstra from the RFS.

On 19 December, further widespread Severe Fire Danger saw significant fire activity. I was on a strike team that initially responded from Sydney to Ulladulla on the south coast but was urgently redirected back to the Green Wattle Creek fire where homes were burning in the township of Balmoral and other locations. Tony Abbott was again out among the flames on another truck in our strike team. We worked on the fire close to the township of Bargo and on a number of occasions were outflanked by spot fires.

One Northern Beaches crew was overrun by fire and their hoses burnt through as they sheltered inside the truck cabin.

As we were leaving that night on a coach following about fourteen hours of driving and fighting fires, we received the news that a tree had fallen, hit a tanker, and killed two fellow firefighters. There was scant information available, and we feared that they may have been from the local crews who had just relieved us. The word came through that two young fathers, Geoff Keaton and Andrew O'Dwyer, volunteers with the Horsley Park Brigade in Western Sydney, were the victims. Everyone was devastated. We didn't know them personally, but they were brothers. I remember looking around the coach, seeing vacant stares, black smudged faces with tears running down them. It was a very quiet ride back home, everyone lost in their own thoughts, tired, upset. One of the prominent thoughts, I know, was, *That could have been me.*

Two days later I was back among the flames on a day of Catastrophic Fire Danger. On 21 December 2019, Sydney was choked with thick smoke. It was surreal to see people wearing P2 face masks to filter the smoke, the sun a huge, orange ball in the sky, and the moon at night a deep red. People were always coughing. Firefighters were coughing more than anyone else.

Erris left for work every morning at 6:30 to manage a large aged care facility. She had been very busy as the fires had forced some other nursing homes to be evacuated and, on days of Catastrophic Fire Danger, other parts of the war vets' village near bushland had to be evacuated as a precaution, many residents coming to her facility. A couple of times she stayed at work to help confused evacuees settle in to their temporary accommodation.

I had been detailed to lead a fire crew that day but wasn't needed until 10 am, so when Erris went to work I dressed in a clean uniform and headed to a cafe near the Fire Control Centre for breakfast.

I had a newspaper with me and looked at the front page. Splashed across it was a big photo of the two young fathers, the firefighters killed a couple of days before. Sleep-deprived, upset, one of my known PTSD triggers well and truly activated, I took some deep breaths and turned away from people so that they couldn't see the emotion on my face. As a PTSD sufferer it can be embarrassing when strong emotions are triggered that are hard to control. I pulled myself together as best I could and ordered breakfast and coffee, then went to pay. I'd noticed that the barista had walked over when I was ordering and assumed that she was showing her colleague how to use the cash register.

But the cashier then said to me, 'No need to pay. It's been taken care of.'

'Sorry?' I said, confused.

The barista, who said her name was Flo, leaned over and said, 'It's my shout – thanks for what you and the other firefighters are doing for us. We are so grateful to all of you.'

A sudden random act of kindness. That was it for me, it brought me undone. I thought of the two young dads and what they had sacrificed. I couldn't speak and my eyes filled with tears. I was able to choke out, 'That is *so* lovely and thoughtful of you. Thank you so much.' Then Flo started to cry! It caused quite a scene. I always make a point of going to say hi to Flo whenever I go to that particular café, and made sure that I introduced Erris to her.

I have always been touched by how people genuinely want to do whatever they can to show they care, to help others. Not everyone can be on the front line. Being a firefighter had its challenges and long-term costs, but one of the huge benefits was seeing the best come out in people.

My strike team of five fire trucks was assigned to the fires at Blackheath in the Blue Mountains. Two other strike teams from

our area headed to Lithgow and Bilpin respectively, in all a total
of fifteen fire trucks and about seventy firefighters. Thousands of
firefighters from the RFS, Fire & Rescue NSW, National Parks &
Wildlife Service, the Forestry Corporation of NSW, and interstate
fire services worked on dozens of fires that day.

We made the two-hour journey with flashing red-and-blue
lights, siren blaring, driving west on the freeway with an assortment of other fire trucks and command cars all heading towards the
ominous smoke plumes in the distance. As officer in charge I was
in the front passenger seat, with the crew behind me. Conversation
was difficult because of the sirens and the engine noise, but we discussed what we were seeing – trees swaying in the strengthening
wind and the temperature outside obviously climbing, eventually
reaching more than 47°C.

There were huge pulsing columns of grey, white, brown and
black smoke on the western horizon ahead of us, to our north
and to our south. Sometimes we would go quiet, lost in our own
thoughts, wondering what the day would hold, if we would be safe,
if we would lose houses to the flames, or if we might just 'hurry up
and wait', as sometimes happens at big fires when you're held back
in reserve.

The smoke to the south was coming from the Green Wattle Creek
fire, to the north from the Gospers Mountain fire that would rage
out of control and impact many homes that afternoon, and ahead
of us, from the Ruined Castle fire south of Katoomba and the Grose
Valley fire near Blackheath, our destination.

The crew was a mixed bunch, including two brand new recruits:
Vicky, who'd been a paramedic in the UK, and Nicholas, a teenager
who had never been to a large fire before. We also had a deputy
captain, Rachel, who was driving the truck and who had more than
twenty years of experience, and two other volunteers, Mark and

Leo, each with about seven years under their belts. I briefed the crew about what we might face and how we would work. I paired up the new recruits with the more experienced members and told them that we always worked to the buddy system – never alone – and that we always had to stay in radio contact.

There was quite a bit of nervous chatter initially, but as we drew closer, the talk died off. I knew the area we were headed to but didn't have an appreciation of exactly where the fire was. I was worried about the wind. As expected, it was strengthening as we climbed higher into the mountains and I knew that a wind change was forecast later in the day.

As we reached the upper mountains, I saw that at least one of the huge smoke plumes reaching high into the sky was starting to look like a storm cloud on top, a bad sign. I snapped a couple of photos and texted them to a colleague who I knew was working in the State Operations Centre. A short time later, everybody's phones pinged with an emergency warning stating that multiple fires were showing signs of pyroconvective activity, and that conditions were becoming dangerous and unpredictable. I later found out that the meteorologists were crunching numbers and radar data that suggested firestorms might be forming, and my photos were the clincher, providing the confirmation they needed.

When we arrived at Blackheath we could see that while there was a lot of fire burning in the valleys, at that stage they were not close to homes. There was no way of safely accessing the bush to get to any of the fire fronts, which were raging through rugged, mountainous terrain. Thick black smoke reached into the sky across multiple fronts and flames were leaping above the treetops, but for the moment no properties were being threatened. We were in defensive mode because of the weather conditions, as any attempt to directly confront the fire fronts would have been dangerous and futile. It was quite frustrating, and I had to explain to the new firefighters that they would become accustomed to the concept of 'hurry up and wait'.

Thankfully, at that stage, the strong wind was driving the fire away from towns and into the Grose Valley, the site of many major bushfires that had caused loss of life and property over the past 150 years, because its southern side was studded with towns along the top of a ridge that ran east, with steep wooded slopes below. Periodically, over the years, massive fires had charged up those slopes and consumed hundreds of buildings, sometimes taking lives as well. They usually came through about once a decade, but more recently the area had been subjected to big fires in 2006, 2013, and now, in 2019 as the frequency of serious fire weather events increased due to climate change.

My father had fought fires in these mountains in 1944 and 1952, and I had fought fires here in 1977, 1994 and 2001. I worked here as Blue Mountains Hazard Reduction Officer back in 1989, and had been up here as commissioner when fires burned through the same valleys in November 2006, thankfully not reaching homes, then again in 2013. I had a good grasp of the risks we were facing.

Of chief concern was a forecast cool change that would bring sustained strong southerly winds but no rain. The southerly would turn the long, less intense flanks of the fire into raging fire fronts when it arrived. As we drove around and familiarised ourselves with the terrain, I spoke to Pete Duff, who was in charge of our other truck, and Deputy Group Captain Sean McLoughlin, our strike team leader. I told them I didn't like what I was seeing, that it wasn't safe. There was a lot of fire to the south of homes on a ridge top with lots of unburnt bush in between, and the southerly wind would push intense flames directly onto them. They agreed, but there wasn't much we could do except watch, wait and prepare.

The command team were obviously aware of the situation and had spread out firefighting resources in preparation, but couldn't work miracles.

Looking to the north-west and the north-east, we realised that we were in a better position than the other strike teams from our home turf that had been sent to the towns of Lithgow and Bilpin.

Bilpin looked like it was ground zero. An enormous storm cloud, growing like a cancer out of the towering smoke plume, loomed above. A fire-generated storm is something to fear because it can lead to very strong, very dry upper-level winds being drawn down to the surface, as the storm cloud links the fire with the upper atmosphere. This leads to updraughts, downdraughts, fire tornados, winds from every direction, and even lightning sparking new fires up to 30 kilometres away. Sometimes, but usually not, they can be accompanied by some light rain, but not enough to slow an out of control fire.

Once rare, with only sixty recorded events of pyroconvective storms in Australia from 1978 to 2018, the climate change fuelled Black Summer spawned at least twenty-nine firestorms.[39] This represents a frightening escalation of a phenomenon that overwhelms firefighting efforts, destroys buildings and kills people. Even from 20 or more kilometres away, the huge smoke column that reached more than 10 kilometres into the sky was visibly twisting and looking as if it was boiling. Further west in Lithgow they were also losing homes as fires raged out of control there.

We later saw television news footage of what colleagues were dealing with and why they lost buildings. The fire behaviour was extreme and the wind conditions, driven by the storm above them, were intense. Gale-force winds were driving a blizzard of hot embers that set buildings alight, set some fire trucks alight, and started new fires ahead of the main fire front.

Firefighting tactics follow a logical progression from extinguish, contain, protect property, to protect life only. Firefighters to our north were onto the last tactical priority and the lives they were trying to save were often their own.

We didn't realise at the time that we would soon be in a similar position. I don't know exactly when the southerly wind change arrived, but conditions altered abruptly. We were on Hat Hill Road, a long, winding, narrow road surrounded by thick eucalypt forest, desiccated by drought and primed by days of high temperatures, dry winds and very low humidity. Dad had taught me in the early days that southerly wind changes on the New South Wales east coast were to be both welcomed and feared. Welcomed because they brought lower temperatures and higher relative humidity (making for cooler and damper conditions), but feared because they were typically very strong at the end of a hot day, and vegetation could take hours to absorb the increased moisture in the air, resulting in fire behaviour remaining extreme for an hour or two after the change. He told me, and I learned through experience over the years, that a strong southerly could make fire run as if it was still hot and dry for two or more hours until the increased moisture in the air took effect, and that the former long northern flank of the fire would suddenly become its raging head. The increased size of the new fire front often overwhelmed firefighters.

That's exactly what happened on 21 December 2019. It was near dusk and suddenly the trees bent over as the strong southerly wind arrived. The sky above went dark as black and brown smoke changed direction above us and obscured the setting sun, then the sky went orange, reflecting huge, towering flames that were racing directly at us. There was a roaring sound like jets taking off at Sydney Airport.

A couple of Fire & Rescue and local RFS crews had set themselves up to protect homes about a kilometre down Hat Hill Road from where my truck was located. When we were down there earlier I had commented on the geography, which would place residents and firefighters at great risk when the southerly arrived. One of the houses was at the top of a small, steep hill, with lots of dry bush between it and the fire front to the south.

Within 10 minutes of the wind shift, the fire units there were transmitting urgent Red messages over the radio saying that they were being overrun by fire and were sheltering inside their trucks with cabin protection sprays on. We headed towards them but there was no way we could get through – flames were roaring across the narrow road along a stretch of about 200 metres, and anyone or anything that tried to pass would have been incinerated. Flame lengths were at least 20–30 metres.

Luckily, we found somewhere to turn around, but we saw that the road was about to be cut off by fire behind us as well. We went into a driveway and found three men preparing as best they could to protect their beautiful old homestead, which thankfully was surrounded by lawns and ornamental trees that were not particularly flammable. We could hear radio messages from the other Terrey Hills truck, which was protecting some isolated homes a few hundred metres further down the road. A third truck with a Terrey Hills crew pulled in behind us and we started to run out hoses between the house and the fire as flames tore through the treetops behind the property.

I took Vicky with me, left Rachel in charge, and went back to the road. It was an eerie scene. Everything was lit up by the orange glow as flames roared above the treetops, bearing down on the road about 200 metres from us. The fire had already jumped the road to our west and our east, but we were fairly safe near the house because of the cleared grounds around it. The fire would soon reach the bush opposite us and the house come under ember attack.

We spent the next few hours in that general locality and joined up with Pete and the crew on our other truck. We checked on the crews that had been overrun by fire. Thankfully they had escaped unscathed. Almost miraculously, there had been no loss of property where they had been overrun, which is testament to the dogged determination and bravery of those firefighters.

We were able to contain one edge of the fire by lighting a tactical backburn from a fire trail. From an early age, Dad had shown me how to get the main fire to draw a backburn into it, rather than be driven by the wind towards you and possibly jumping the control line. As Dad had shown me many times, it is all in the timing. GoPro footage of the fire shows the swirling wind and flames, and at one stage I call out to the crew, telling them to run to safety. Luckily, though, I was able to light a few spots along the trail at precisely the right time, and they were then sucked in towards the main fire against the prevailing wind, rather than jumping to the unburnt bush behind us, threatening more houses, but worse, potentially trapping us while we were on foot with no protection.

It must have been after 11 pm, more than fourteen hours after we'd started the day, that we finally packed up at that location and thought we would soon start the long drive home. However, we were immediately redirected to a burning house nearby. We extinguished the burning carport, then once again thought, *Time to go home.*

Driving over the crest of a hill to an intersection, the sky suddenly turned orange ahead of us. Flames had crept around the bottom of a deep gully in a sheltered location and then come under the influence of the still-strong southerly wind. There had been very little moisture recovery and the humidity was still very low. Firestorm is an apt description of what we saw as we crested the hill. Flames were roaring up a steeply wooded slope directly behind a long row of houses on Hat Hill Road. A couple of the houses appeared to be on fire and radio messages suggested that this indeed was the case.

There must have been about ten fire trucks from the RFS and Fire & Rescue, and firefighters were desperately running out hoses from their trucks and from hydrants down into backyards as quickly as they could. Footage of one of the Terrey Hills crews taken at that stage went viral around the world, showing flames dwarfing a two-storey house that was being blasted by an ember storm. Equipment on the back of their truck caught fire but luckily the crew saw it in time and put it out.

I realised that firefighters wearing bushfire fighting uniforms would not be able to safely work with their backs to the houses, as they would be engulfed in heat, smoke, embers and flames. Bushfire fighting uniforms have to be light enough to keep firefighters cool during hot weather, and thick enough to protect from a reasonable amount of radiant heat. This fire was too hot for firefighters to put themselves in backyards between homes and the flames, just metres away. After speaking to Pete, who ran up to us as we pulled in, I asked one of the more experienced members, Mark, to join me in putting on his thicker, more protective structural firefighting uniform and a self-contained breathing apparatus set. Rachel took charge of the rest of the crew again and they went to protect another house.

Mark and I then went down the front path of a house that was being blasted by sparks and flames, and met two firefighters who were being beaten back, trying to shelter around the side of the house while still bringing their hose streams to bear as best they could. It was a losing battle and the hose streams were ineffective against the gusting wind. We grabbed the hoses off them and went through the heavily overgrown backyard to make a stand as the fire blasted up at us out of the steep gully.

I remember reflecting that I could see that we were being enveloped in flames, but I couldn't feel anything through the thick protective uniforms that totally enclosed me and Mark. The smoke couldn't affect us as we had about 30–45 minutes of fresh cool air in the cylinders on our backs. We directed our hose jets into the bush about a metre below us and eventually, probably minutes later but it seemed much longer, the intense flames died down.

Inexperienced firefighters often direct their hose jets upwards at the flames, thinking this will have an effect, transfixed by the huge flames looming above them. This never works. The water simply evaporates and makes no difference to the intensity of the fire. You must always attack the base of the flames and try to bring the fuel below its ignition temperature. Moving the hose jet around too much, what I call garden hosing, also has little effect. You need to concentrate on a particular point until it cools then move the water jet on to another spot. This can be unnerving as it takes a little while to have any effect, but you have to stay the course.

I was confused when I lost vision a few times through my mask, finally realising that each of the multitude of embers engulfing us were small pieces of burning bark, grass and leaves. We would put the nozzle on a spray pattern when we were being enveloped in flames and embers, and our masks would quickly become covered in a layer of wet, black ash, like black mud, so we had to constantly wipe our gloved hands across our masks in order to see again.

Mark and I went from house to house, taking over hose lines from less well-protected firefighters, knocking down the flames, handing the hoses back, then moving on, until our air supply was close to running out.

At one stage a Fire & Rescue NSW firefighter, reading my surname on the back of my helmet, said, 'Hi, sir! So . . . how are you enjoying your *retirement*?' We all got a bit of a laugh out of that.

Eventually, some time well after midnight, we were finally released to return to our fire station back in Sydney. On the drive back I was freezing cold, my head was aching and my kidneys were sore. I knew that despite drinking bottles of water I was probably dehydrated. The sun was rising when we eventually pulled up after a two-hour trip. Before going home, we had to service our breathing apparatus sets, put dry hoses on the trucks, restock them and ensure they were ready, as fresh crews were heading out again later than morning. It had become a familiar drill over the months of Black Summer.

As I fell into bed some twenty-five hours after leaving, I reflected that it had been some of the most intense firefighting that I had ever experienced, and also, reluctantly, that I am not as young as I used to be. Erris, a registered nurse, was not particularly happy with me when she arrived home from work later that day, as she suspected I was suffering from bad dehydration. She told me that I should have gone to a hospital or sought out paramedics at the fire to administer some intravenous fluids, but I'd left it too late. Instead, I rested for a few days, two of them in bed, and drank as much water as I could. The dull ache in my kidneys slowly got better.

On New Year's Eve, teams of volunteers were detailed to be at the Northern Beaches RFS Fire Control Centre at 5 am to set out in a convoy of bushfire tankers to the far south coast and reinforce local fire crews on what was forecast to be yet another day of serious fire weather. Many fires burned in the vast forests and mountains behind south-coast towns, and as had happened in 1968, it was predicted that today they would forge their way to the coast on the back of strong westerly winds. The forecast was for Extreme Fire Danger.

I had heard on the news the night before that a young volunteer fighting a fire on the New South Wales / Victorian border,

Samuel McPaul, had been killed when a downdraught of strong winds from above, possibly from a pyroconvective storm, picked up his fire truck and dumped it on its roof, killing him instantly.[40] The rest of the crew received burns and other injuries as the fire seemed to explode around them, fanned by a cyclonic wind burst. I learned later than one of the burn victims had to wait two hours for medical attention and pain relief because the fire prevented help from reaching him.

The fire weather forecast for New Year's Eve, and the fact that a large number of fires were burning in close proximity in rugged high country, made me think that more fire-generated storms were likely. Research has established that multiple large fires in close proximity in steep, rough country can provide some of the ingredients needed to create conditions conducive to pyroconvective storms, as seen in 2003 in Canberra and 2009 in Victoria. Proximity and the heat generated by fires burning intensely up steep inclines make it more likely that convection columns will interact and reinforce each other, causing extreme fire behaviour and wind effects, and enabling the heat to reach into the stratosphere. Given these preconditions I knew that pyroconvective storms generating lightning and high winds from varied directions were therefore highly likely.

As I had often done before, I pondered what the bloody hell I could do to keep my crew safe if we were caught underneath one. There was no answer to that particular hypothetical, and all it elicited was a shudder down my spine. This was something that I hadn't had to worry about in my early years of firefighting, because the majority of firefighters had never actually seen a pyroconvective storm. That is, not until this horrific summer when they became almost commonplace on the worst days.

Two strike teams, each comprising five 3,400-litre four-wheel drive tankers crewed by five firefighters as well as a four-wheel drive command vehicle with a crew of two, were to drive to Batemans Bay on the far south coast of New South Wales – a five-hour trip. Every other Rural Fire Service district in Sydney and from many other areas of the state were also sending strike teams, as was Fire & Rescue NSW. In addition to these agencies, the NSW National Parks and Wildlife Service and Forestry Corporation of NSW were sending more resources to the area.

We set out at about 5:30 am. I checked the RFS Fires Near Me app and saw that already there were several Emergency Warning and Watch and Act alerts in place for fires to the west of towns on the south coast, as strong westerly winds started to blow – not a good sign given how early it was in the morning. The maps showing the perimeters of the fires told a sobering story. Strong winds had been blowing throughout the night, fires had moved many kilometres towards coastal townships and fire perimeters were now immense and totally uncontrollable. On checking the Bureau of Meteorology's bushfire weather page, I thought to myself, *Oh shit! Here we go.* It was going to be a very, very bad day.

I thought back to earlier years when we didn't have ready access to such tools, to when we simply used to wing it with limited infor-mation in a much simpler world when weather patterns were fairly predictable and when prior experience and local knowledge was a fair guide to what to expect. I realised that those simple days were another casualty of climate change – we would be in uncharted territory during the worst fire seasons from now on.

Things certainly weren't good. Temperatures were increas-ing rapidly, relative humidity was dropping and strong winds had been blowing for hours. This meant that we would face a full day of horrendous fire weather and fire would be able to cover huge dis-tances. One of the main considerations when planning for fires is how many hours of serious fire weather to expect. The earlier in the day bad fire weather starts, the less chance of controlling existing fires and greater the chance that there will be fresh outbreaks, often caused by power lines clashing and arcing in strong winds. For years I had advocated for the undergrounding of power lines due to the increasing intensity of storms and increasing wind velocities during bad fire weather, making power infrastructure and communities increasingly vulnerable. This, however, would be a massive, expensive undertaking. It underlines the rising costs involved in adapting to our changing climate, and the economic consequences of continuing to take insufficient action to try to halt and then drive down warming.

The fire danger forecast resulted in the premier, under State of Emergency powers, ordering an unprecedented evacuation of hundreds of thousands of holidaymakers from the popular holiday destinations that were packed due to the summer school holidays. A big call, but a very good one, which I believe saved dozens or perhaps hundreds of lives that day. Premier Berejiklian was showing her mettle as a great leader, unlike the embattled PM who had been forced to return to Australia from an overseas holiday after a community and media outcry.

We had a great crew. Andy was the driver, a deputy captain and veteran of more than a decade of fighting fires. Jessica, a nurse, who had grown up in the RFS with her father, a former brigade captain, and her brother – Jess was very experienced and held many qualifications. Tom, who had been a member for a couple of years and had joined with his dad, Sam. And Matt, a solid, experienced firefighter and brigade training officer. Pete Duff, our captain, was on another truck in the same strike team.

Soon after setting out, as we approached the Sydney CBD, I heard a radio message: 'FIRECOM to Northern Beaches strike team leaders Alpha and Bravo, please instruct all of your appliances to immediately respond and expedite to Batemans Bay.' I acknowledged the instruction, nodded to Andy, then turned on the flashing lights and siren.

As we passed through Sutherland, on the southern outskirts of Sydney, another two strike teams, ten tankers and two command four-wheel drives from another district merged onto the highway, red-and-blue lights also flashing and sirens blaring, forcing their way through the traffic as cars pulled over to let the stream of emergency vehicles through. Soon after, a squad of four fire engines and a command car from Fire & Rescue NSW overtook us. The urban fire engines had the advantage of bigger engines and smaller water tanks, making them lighter, and would reach Bateman's Bay before us.

I realised that there were probably more than 100 fire trucks, carrying hundreds of firefighters, heading to the south coast at that moment, with many more to follow, and many already out fighting the flames. It must have looked like Armageddon to people travelling in the other direction. For many that day, it was, as thousands would be forced onto the beaches and into the sea to save their lives. Sadly, some would not make it. Hundreds would lose their homes and their livelihoods.

I knew that interstate assistance had been requested, but unfortunately other states were having their own fire problems, restricting what they could send. Gippsland in Victoria to our south was experiencing the same fire weather conditions with many large fires already burning, and South Australia was also being hit hard.

This is another common feature of the new normal of climate change – simultaneous fire seasons, restricting the previous practice of routine cross-border sharing, possible in the past when fire seasons had been sequential, with little overlap. As pointed out in several of the Inquiry reports, this is now a given: that there will be increasingly

limited ability to share firefighting resources across borders in future
as the climate becomes warmer and even more dangerous. And of
course, our fire seasons are also now increasingly overlapping with
those in the northern hemisphere, restricting access to the large firefighting aircraft that we lease from Canada and the USA.

A few hours later, our convoy of fire trucks drove through the
twin towns of Milton and Ulladulla, sirens still blaring to clear
the traffic. As we left Ulladulla, still heading south, I looked
inland to the west. My blood ran cold. Treetops were swaying and
bending towards the ocean due to the strengthening wind, and it
was already hot at about 8:30 am. I could see in the distance, the
closest probably 20 kilometres away, several big smoke plumes,
grey and brown turning black and bending towards the east as
strengthening winds pushed smoke, flames and burning embers in
the direction of what firefighters often called the 'great eastern fire
break' – the Pacific Ocean.

But the smoke plumes, wind and heat were not what chilled me
most. As we desperately raced towards the conflagrations that we
couldn't yet see, we passed a stationary line of traffic: thousands of
evacuees trying to escape north to safety. Families who had been
holidaying and camping, many towing trailers, boats and caravans,
were all trying to flee, abiding by the urgent evacuation order.
I looked at the smoke plumes and trees bending in the wind and
thought to myself that the police probably had about three hours,
possibly four, to clear the highway before the flames swept across,
consuming everything in their path.

Fires that could generate 60-metre-high flames and fling burning
embers up to 12 kilometres ahead of the main fire to start new blazes
wouldn't even slow for a moment on reaching a two-lane highway.
The cars would simply provide additional fuel. I couldn't imagine
how the police could do it, and my heart lurched.

The traffic jam went for many, many kilometres and all I could think of was a horrifying image indelibly imprinted on my brain during my childhood, in about 1965: news footage of a line of burnt cars on a freeway near Avalon in Victoria, with burnt bodies covered by blankets on the road, the blankets flapping in the strong wind. The motorists had been killed when a fast-moving bushfire had overtaken them, regardless of whether they ran or tried to shelter in their cars. One of the golden rules in bushfire education is 'never be caught on foot or in a car during a bushfire – you will not survive'. What I saw horrified me and I had a terrible sense of foreboding that made me feel ill.

The scene as we finally reached the town of Batemans Bay was surreal. Gale-force wind gusts buffeted our truck. The temperature was already in the mid-30s. The smoke ahead of us as we crossed the bridge into the town centre was black, brown and grey, but mainly a strange orange colour.

It was obvious that, over the lip of a small hill just beyond the bridge that funnelled traffic into the main shopping area, the fire was crowning, burning through the treetops, which I knew in that area were commonly 30 metres high. It was only about 10:30 am, and homes on the outskirts of the town were already ablaze, a scenario playing out at many locations across a 200-kilometre stretch of coastline in New South Wales and to the south in Gippsland, Victoria.

We could see flames up ahead and a column of thick, black smoke, obviously coming from a large, burning building, but we were ordered to head to suburbs being heavily impacted by fires where many homes were threatened or burning.

The next twelve hours or more were a blur. Our first stop was Surf Beach, where we teamed up with crews from Fire & Rescue to save a number of homes. An off-duty ACT Fire & Rescue firefighter ran up to us as we arrived and connected a hose to our truck, racing back to save his own home as his wife and children sheltered inside, terrified. We relocated often, at one stage working hard to stop flames and radiant heat from a burning home setting houses on either side alight, Matt and I in breathing apparatus until our air supply ran out. Pete Duff and our other truck were dealing with yet another burning home about 50 metres away from us. In a neighbouring street we beat back flames but were unable to save an expensive boat on a trailer. Just as we packed up to leave we heard a loud crack, and a 15-metre-high tree crashed to the ground next to us, blocking our exit – a near-miss.

We cut up the tree with our chainsaw, then responded to the suburb of Catalina. Many homes were burning, and we could hear loud screeching noises as pressure relief valves on large gas cylinders attached to the external walls of burning homes vented, shooting out plumes of flame as the venting gas burned. We heard explosion after explosion and saw mushroom clouds of flame through the dense smoke as gas cylinders tore open after red-hot steel cylinders lost their strength and the boiling liquid inside was released. As the highly flammable LPG escaped and instantly changed state from liquid to gas, it expanded 270 times and immediately exploded. The technical term is boiling liquid expanding vapour explosion, or BLEVE. A BLEVE is something you don't want to be anywhere near, so we steered clear of the sound of venting cylinders going off like rockets, and also of the many fallen power lines that could also kill.

There was nothing we could do for most of these homes but we found some that were undamaged and others that had decking and other parts of the house on fire, which we were able to quickly douse. In one location flames had run across people's mown lawns and set fire to garden beds, which then set fire to the houses.

On the western side of the Batemans Bay CBD it was surreal to see a large two-storey electrical store and warehouse burning from end to end, with no firefighters present. About 100 metres down the road, a wrecking yard full of cars, an office and a warehouse were blazing. As we pulled into a clearing at the top of the hill, I saw the local State Emergency Service headquarters burning fiercely, the metal roof glowing red hot and black smoke pulsing out. There was absolutely nothing that we could do and, like thousands of other firefighters that day, at times I felt utterly useless and powerless.

We had been going from house to house and between towns and suburbs doing whatever we could, but often the fast-moving fire fronts had already charged through, leaving destruction in their wake. On too many occasions, by the time we arrived, homes were well alight and beyond saving. But, in hindsight, we were able to save many. Unfortunately, firefighters tend to focus on what is lost, not on what is saved, and that can have a serious impact on mental health. I tried to stay positive, but when you go into an area where most homes are already well alight, it's hard to find anything positive to think about.

The toll of homes destroyed in that fire alone on New Year's Eve, the Currowan / Clyde Mountain fire, was 490. Hundreds of other buildings like shops, warehouses and sheds were also destroyed. Throughout December and January, around 1,500 homes and 3,400 other buildings were destroyed in the south of the state, most of them on New Year's Eve. Compared to our previous worst property loss fires in New South Wales in 1994 and 2013, losses in 2019–20 represented an increase of more than 1,000 per cent.

Since New Year's Eve I've had intrusive memories of one particular house. As we slowly drove past, I asked Andy to stop. I remember we could see through the glass windows that a single lounge chair was burning, and I thought, 'We can save this.'

The other crew members pointed out something that I hadn't initially seen because of all the smoke in the air – that the roof was well alight – a very dangerous situation because if we entered to fight the fire directly, the blazing ceiling and timbers could, and at some stage would, collapse in on us. The house was already essentially lost, but my mind continues to play games and just focuses on that single burning chair. What if . . . If only . . . I know that thousands of other firefighters and fire victims throughout Australia still grapple with their own versions of the burning chair, and much worse, particularly those who faced threats to their own lives and those of loved ones. When I used to train new recruits and aspiring officers back in my days in the Bushfire Section of the NSW Fire Brigades (now Fire & Rescue NSW), I would always warn about the first major bushfire where they would face multiple losses of homes, and maybe even deaths. Such situations are very hard to process and accept for somebody whose profession is the saving of life, property, and the environment. I suppose it could be a form of survivor guilt. Whatever it is, it takes a significant toll.

On a day that never seemed to end, we were asked to deal with a reignition in a small portion of unburnt bush in the historic township of Mogo. It had been devastated and much of the main street lined with historic buildings lay in smoking ruins. One sight will stay with me forever. From my experience, fast-moving animals such as kangaroos generally seem to know where to go to find safety during a fire, whether it is jumping through a section of low flames onto burnt ground, into a creek or dam, or simply outrunning the flames.

But what I saw as we prepared to head to Mogo made my stomach lurch and my eyes fill with tears. A number of black, smouldering shapes next to the highway. At first thinking they were burnt hay bales, I tried to focus through the smoke, then noticed some of them twitching.

Animals have strong survival instincts in the face of bushfire. Over tens of thousands of years of bushfire seasons, native species have learned and passed on a multitude of different survival strategies. They fly, run, hop, scramble or burrow to escape fires. At times, they'll even turn to humans for help.

There was a time, back in the summer of 1994, when I had set up a command point in the car park of a sporting oval. As I got out maps and portable radios, I saw two ringtail possums the size of small cats scuttling across the ground towards me. I thought that they mustn't be able to see me and ignored them, thinking they would head for unburnt bush behind me.

Next thing I felt a tugging at my leg and one of them climbed me like a tree, perching itself on my shoulder. I put my hand up towards him thinking 'Ohhh, how cute!', and for my efforts received a firm bite on my finger that drew a bit of blood. Next thing the other possum climbed up too, and there I was with a possum on each shoulder. This big brave firefighter wasn't brave enough to try my luck again, and I just stood there wondering what the hell to do with a man-eating possum perched on each shoulder!

A resident walked over, trying to stifle his laughter, and asked what I was going to do. I wasn't game to try to dislodge them again. I figured they felt secure on my shoulder and wanted to get away from the fire. I walked about 100 metres to unburnt bush and leant my shoulder against a tree branch. One of the freeloaders immediately hopped onto the branch. I then turned my other shoulder to the branch. Nope. Not moving. This was the nasty one that had bitten me. The resident, who was laughing heartily by then, handed me a stick. I tried to prod Mr Possum, but he bit the stick and dug into my shoulder with his sharp claws. Ouch! Finally, with dignity and aplomb, making utterly sure that I knew exactly who was in charge, he climbed off, taking his time about it, then the two possums disappeared into the bush, safe from the fire.

At the time it reminded me of a story Dad often told about big fires near our home, before I was born, in 1957. While cutting a trail and putting in a backburn, a bandicoot had scurried out of the burning bush, raced up the handle of Dad's rake, then sat, quivering with fear, on his shoulder. He wouldn't move until Dad walked back up the track to an unburnt area where he allowed Dad to lift him off and put him on the ground. Dad said the bandicoot sniffed his boot, then scurried away. Sometimes I think animals know instinctively that (most) humans will help them.

Now, in the Black Summer of 2020, nothing could have helped these poor animals going through their death throes in the choking smoke. It was a small mob of about six kangaroos that had been overrun by flames. It appeared they had run as far as they could through the blazing bush, then dropped dead, literally burning, on the road. A couple of weeks earlier at the Three Mile fire, I'd seen mobs of kangaroos bounding across the head of the fire to safety – they knew where to go.

But I'd never seen anything like this before, and hope I never do again. It told me that the speed, size and ferocity of the fire left no escape for what would later be tallied as up to 3 billion native animals killed or displaced by the biggest forest fires ever recorded in Australia. As the months passed I became more, rather than less, upset by what I'd seen. I had a dawning realisation about the connections between: the widespread devastation of the natural environment, both flora and fauna; my efforts to get facts about climate and fires through the cloud of disinformation; and my feelings of impotence against the massive forces of nature being unleashed as a result of our continuing addiction to fossil fuels.

I later realised that I would probably never again in my lifetime see the teeming wildlife I had seen as a child when camping with my parents, and that my grandchildren would never experience the wonders of nature that I did as a child. It will take decades or even centuries for the environment to recover from the Black Summer fires, if a full recovery is even possible given the increasing frequency and intensity of fires.

While I was fighting fires at Batemans Bay and Mogo, the disaster was playing out at many other locations on the New South Wales and Victorian coast. On the way to Batemans Bay I had passed the turnoff to Lake Conjola where my ELCA colleague, Peter Dunn, lived with his wife Lindy.

Following a distinguished military career, Major General Peter Dunn was appointed commissioner of the Australian Capital Territory Emergency Services Authority and given the task of restructuring Canberra's fire and emergency services after the devastating 2003 bushfire disaster.

In early 2019 he joined ELCA, deeply concerned about worsening natural disasters driven by climate change. He was soon to be reminded firsthand of the dangers he had tried to warn the Australian Government about, and also of his experiences as a Canberra resident during the 2003 firestorm.

Peter and Lindy had been nervously listening to reports of the growing fires for weeks leading up to New Year's Eve. They were particularly worried about the Currowan fire which had started on 26 November and steadily moved closer and closer to Lake Conjola.

Peter told me that he listened to a radio interview one day where the local head of the RFS expressed his concerns that in the coming days and weeks the fire could burn 'all the way north to Sussex Inlet', meaning that it would reach where they lived.

This spurred Peter and Lindy into action. They spent a week moving flammable materials away from their house, cleared leaves from the roof gutters, bought long garden hoses that reached all around the house, and ensured they had metal hose fittings. They developed an evacuation plan: if the fire hit an area of bush about 150 metres from the house, they would either drive to the beach, or, if that wasn't possible, jump into the tinnie at the bottom of their backyard, and follow the river to the main lake.

They had been comparing RFS predictions of fire spread to what subsequently happened and found them to be reasonably accurate over several weeks. They were relieved on 30 December 2019 when they looked at fire spread predictions suggesting the Currowan fire would not cross the Princes Highway and impact the Conjola community.

On New Year's Eve, Peter and Lindy went for a run. He told me that as they jogged down a bush track towards the lake, they ran into an area of very hot air that made it difficult for them to breathe. This would have been at about 9:00 am. They ran along the beach, had a quick swim, then as they left the water, were horrified at what they saw. Black smoke was billowing above the small settlement of Conjola Park about 6 kilometres to their west. The fire had obviously crossed the highway.

They raced home, got dressed in protective clothing, and put their fire plan into operation at about 11 am. Peter saw by the colour of the smoke that homes must have been burning. They realised that there was now no option to leave – the only road in and out was cut off by fire.

At about 1 pm they saw that two separate fire fronts were bearing down onto the lake, one from the west and one from the north-west.

They could see flames, and then suddenly a small island about 50 metres from their backyard was burning. The heat, noise and wind were intense, and visibility was down to about 20 metres due to the thick, choking smoke.

Peter recalls hearing thunder, looking up and seeing lightning flashing – a pyroconvective storm. A few drops of rain fell, but in the roaring wind it made no difference at all. On the road in front of their house, people were heading towards the beach, many obviously in a panic.

With the bush to the west and south of their house now burning fiercely, the trigger point for their evacuation plan was reached. Peter said they decided to take the car because, 'It probably would have burned if we'd left it.' Approaching the ocean he saw two off-duty Fire & Rescue firefighters and one of their wives, a retired police officer, directing traffic at a makeshift roadblock. Panicking drivers were speeding through a large camping ground where there were thousands of holidaymakers. A serious accident would have been inevitable but for these three taking charge. Peter said they had no official status but people appreciated what they did. They shared bits of uniform: one had on a fire T-shirt, another a Fire & Rescue jacket, and the other Fire & Rescue pants.

There were cars everywhere and people were heading towards the water as flames approached around the lake. Yet another fire front moved in from the south. Peter said, even though he experienced the Canberra bushfires in 2003, he was surprised by how quickly the fires joined up.

There were no emergency services visible in the area. A man in an RFS uniform with Media on the back arrived and got swamped with people asking questions: he was only interested in taking photos and Peter said this 'really pissed people off'. Peter suspected that he was probably from a media organisation, rather than the RFS.

Two helicopters then appeared overhead. He saw that the southern fire was approaching the camping ground and feared that deaths would occur if it was wiped out. The helicopters made a stand, bucketing the fire front relentlessly, saving the campground and people in it as well as the village of Lake Conjola. As the helicopters hovered over deeper parts of the lake to scoop up water, they had to dodge jet-skis and boats speeding down to the beach with people rescued from the west.

Peter and Lindy said that, like everyone, they were really worried. There was nothing they could do but stay by the water and watch as fires raged around them. Thick smoke billowed from the direction of their house and they resigned themselves to finding a blackened, smoking ruin once this was all over. Together with hundreds of others they stayed at the beach until the fire exhausted itself and ran out of fuel.

They were amazed to find their home still standing. Peter said that it was surreal going inside: their feet were wet and they left black footprints on the floors. Everything was covered in a thin layer of ash and soot.

Peter said that he reflected back on the day he had stood with me and other former fire chiefs a couple of weeks earlier at a press conference in Sydney where we tried, yet again, to get the attention of the federal government and warn that we were facing catastrophic fire risks. He felt that he had wasted his time. That the government was totally fixated on media spin and climate-change denial. He said that he felt 'gutted, betrayed', and that maybe he would have been better off spending that time helping his small community prepare.

But he and Lindy had to get to work. The only way out was cut off by fallen trees, and fires still raged in surrounding areas. There was no electricity, no landline or mobile phone reception, and no organised services whatsoever.

On the second day an order was given for all tourists to be evacuated, but the 6-kilometre drive was taking six hours owing to traffic gridlock and periodic closures as fires continued to cut roads. Locals cooked food for the evacuees on gas barbecues, provided drinking water and let people into their homes to use toilets. Many evacuees had to sleep in their cars overnight.

The next problem was a potential public health issue. Food in garbage bins put out for the scheduled New Year's Eve collection started to putrefy. Locals used bolt cutters to break into the local rubbish tip and carted waste in utes and trailers. The next problem was raw sewage. Loss of power meant that household pumping systems didn't work, resulting in some homes experiencing overflows.

What to eat with no refrigeration? Locals were extremely grateful when the volunteer Marine Rescue boat from Ulladulla arrived offshore, packed with tinned food from an IGA supermarket, accompanied by a flotilla of private jet skis and boats, all carrying donated food. Residents formed a human chain to bring the food from the beach to the community centre, which then became a distribution centre. Thankfully, drinking water had been maintained in Lake Conjola by a generator at the water pumping station. Up the road, Conjola Park had no water.

After seven days, there was still no large-scale, coordinated external help. Lake Conjola residents realised that they had to take matters into their own hands. At a gathering of community members, Peter and Lindy were asked to be the volunteer emergency recovery coordinators, a role they fulfilled for ten months.

Peter gave a number of emotional interviews after the fires, clearly traumatised by what they had gone through, but as always he was direct and articulate. He remains frustrated at the well-intentioned but in some cases poorly prepared and coordinated government recovery efforts. The Lake Conjola community received approximately $400,000 in donations and grants for its own recovery, but bureaucratic barriers had prevented any significant expenditure of the funds up until March 2021.

Peter and Lindy made the tough decision to relocate. They said that having to regularly drive through 7 kilometres of blackened wasteland whenever they left home was a constant reminder of the crisis they had faced, and ultimately they realised they had to prioritise their own health and wellbeing over what increasingly seemed to be an endless fight with various bureaucracies. They have now moved to a new home in an area less prone to bushfire threat. They still, however, maintain close ties with their many friends in the Conjola area.

To the north of Lake Conjola, on New Year's Eve, retired Fire & Rescue NSW deputy commissioner and ELCA member Jim Smith fought fires in the Bay and Basin area as a deputy captain of his local volunteer RFS brigade. He later described to me how he and his crew were overrun by fires several times as they concentrated on saving life and trying to steer fires around properties. There was no hope of controlling any of the explosive fires, and by midday pyroconvective storms had formed, making fire behaviour even worse.

ELCA member Bob Conroy was in Batemans Bay in charge of an RFS tanker attached to the other strike team from the Northern Beaches. We saw each other fleetingly a couple of times as we raced from fire scene to fire scene. Bob and I would later compare various stories of that day, and how futile our efforts felt in the face of the extreme weather and intense fires. When we caught up at Batemans Bay RFS station at the end of the day, his truck was covered in pink fire retardant. He explained that a large aerial tanker had dropped its load on an advancing fire front just in the nick of time, allowing them to save a number of homes. The retardant coats vegetation and makes it harder to burn, and Bob's truck was lightly coated by some overspray as the plane flew overhead. It is essentially a fertiliser, so has a low impact on the environment. The colour is added so that lines are easily seen from the air when aircraft are trying to build fire breaks.

The NSW minister for transport and roads and local member of parliament, Andrew Constance, was fighting to save his own home and trying to assist as many community members as possible. He earned great admiration from his community that day and afterwards because of his strong advocacy and practical support to people who had been impacted. He was among a crowd of 1,000 people driven to the beach and into the water at Malua Bay, not far from where I was at the time, to escape walls of flame, falling trees and a gale-driven ember storm. He told me how he didn't think he would make it as he drove through flames, literally fighting for his life. Obviously scarred by his experience, he later told the media, 'We're bloody lucky we didn't bury thousands of people. I'm going to dedicate the rest of my life to making sure this doesn't happen again . . . Certainly it was my first experience of a global event that had a raft of causes but the predominant one was climate change.'[41]

Meanwhile, south of the border, former Chief Officer of Forest Fire Management Victoria and ELCA member Ewan Waller had spent previous days preparing his farm in Gippsland for fire impact as he watched ominous smoke plumes multiplying and getting bigger following multiple lightning strikes in remote areas throughout November.

He had spent his long career as a forester and fire manager, developing a deep understanding of fires in the Victorian High Country.

He knew that today was going to be a crisis, with unprecedented fire weather conditions combining with a huge amount of uncontained fire burning across the landscape. Even without new ignitions it was going to be an immensely challenging day. A day that some would not survive.

Ewan had been a staunch supporter of ELCA since joining earlier in the year. Though he copped criticism from some quarters when his advocacy for climate action had been misinterpreted as a disavowal of managing fuel levels through hazard reduction burning, there was no doubting his resolve in wanting to see action on climate change. Ewan's background in land management was crucial to ELCA, and his expertise in fuel management helped to ensure that we could present comprehensive, nuanced and scientifically valid positions in our various submissions to the Royal Commission and other inquiries that, given the spectrum of views on this subject, sought to represent a middle ground that all ELCA members could live with. He knew, as the Royal Commission would later confirm, that hazard reduction makes negligible difference to fires on days of Extreme and Catastrophic Fire Danger, but also that controlled burning will remain a critical tool in our increasingly limited arsenal of mitigation tools. The grim reality of weather conditions on New Year's Eve 2019 and on the many days of Extreme and Catastrophic Fire Danger throughout Black Summer, however, was that hazard reduction burning could not have prevented such a widespread catastrophe.

Least of all at Mallacoota.

In one of the defining moments of our Black Summer, the Victorian township of Mallacoota, not far from Ewan's farm, was cut off from the rest of the world by fire and about 4,000 people were driven into the water – their only hope of survival – while dozens of homes and other buildings burned.

Eventually, people would be dramatically rescued by the Royal Australian Navy, using ships and helicopters. There was no other way out of the area because roads were blocked by fallen trees, blocking deliveries of essential fuel and other supplies. The flames had also taken down power lines, so there was no power for water, sewerage, refrigeration, lighting, air conditioning, communication or cooking. The only option was to get people out of there or eventually they could have starved. All this in a wealthy, developed country in the twenty-first century.

Along the 200-kilometre stretch of devastation, some localities had no essential services for weeks, with food and water deliveries unable to get through blocked highways framed by a smouldering, silent, lifeless wasteland. No birdsong, no scampering animals. Just death and destruction on a massive scale.

As New Year's Day ended (though it was hard to tell, since it had been as dark as night since lunchtime), we were told that we could not head back to Sydney due to the blocked roads and fallen power lines. The drivers of the fire trucks were exhausted, and we would have refused to go if they'd tried to send us. By then we had been on the road and in the smoke and flames for about fourteen hours.

We were hungry, thirsty, filthy, dehydrated and utterly exhausted. Everybody was coughing as if we had been lifelong smokers, with the bushfire smoke reducing visibility to about 150 metres. There was an eerie orange tinge to the approaching night, and a strange silence. I remember watching bats wheeling in the sky during the day that had turned to night, their flight frantic and confused. It was surreal, and neither humans nor animals could make any sense of the chaos.

We had no power, no streetlights, no mobile phone coverage and therefore no internet, mapping, or emergency warning information. The lack of power had a knock-on effect, with reduced water pressure, no hot water, failure of sewage pumping stations, loss of refrigeration, loss of ATMs, and an inability to get fuel from petrol stations.

We went to the local RFS station at Batemans Bay, where around twenty fire trucks were parked. Local firefighters and spontaneous volunteers from the community had decided that we were all going to be fed – somehow. They had a gas barbeque working, and volunteers were buttering bread rolls, cutting up tomatoes and cucumbers, and making salads. A ute arrived and a couple of strong blokes manhandled another barbecue off the back and set it up. Soon they were churning out snag rolls as well. To us it was gourmet fare. But supplies soon ran low and dozens more needed to be fed.

Someone arrived with large parcels of sausages, explaining that a local butcher had donated them. His refrigeration had no backup power supply, so he wanted to make sure they were put to good use. He couldn't think of any better use than feeding the hundreds of firefighters, as well as a number of people who sat, obviously in shock, staring at nothing. The newly homeless.

As I lay on my back on the side of the road with my eyes closed, trying to process some of what we'd been through, I was almost moved to tears when a young girl, about eleven years old, came up to me and gently said, 'Excuse me, are you awake? Here, have a bottle of water. Thanks for coming to help us.'

All of a sudden it felt like I had a big football stuck in my throat. I gave her a smile and choked out a 'Thank you'. I didn't trust myself to try to say anything more and had a bit of a tear in my eye, battling strong emotions. Off she went, handing out more drinks to weary firefighters.

I have seen throughout my career that adversity, accidents and disasters seem to bring out the best in the majority of people. People instinctively want to help and to show that they care. They want to say thank you to those who put themselves in danger on their behalf. I have always been touched by this and it gives me hope for the future. If we truly care about future generations and about the environment, we will all fight hard for climate action so that they are less likely to face the same kind of ruin that lay around us that day.

In the following weeks and months there would be a worldwide outpouring of shock, sympathy and solidarity. People from Australia and overseas would donate millions of dollars to help victims and to assist volunteer firefighters. This to me was a reaffirmation of the basic goodness and kindness of the human race.

After being fed, the next problem was accommodation. It was a logistical nightmare. Local accommodation had mostly closed because of the loss of electrical power. Some evacuees and newly homeless had been taken into motels and other accommodation. There was no way any of us wanted to be prioritised over them. If need be, we would just sleep on the side of the road.

I received word that we would probably have to sleep on the floor of the RFS training centre at Mogo. As long as we had a blanket and a roof over our heads, we didn't care. We were exhausted and needed sleep, but the Mogo plan fell through. The fire had gone through the training ground and plastic sewer pipes from the toilets had melted, rendering them and the showers unusable and unsafe.

At around 9 pm, Wayne Reeve, an RFS Inspector from the Northern Beaches who had come all the way from Sydney to try to ensure we were looked after, found a few motels with empty rooms willing to take in firefighters. It was an incredible relief to go into a room, have a (cold) shower, then immediately fall asleep between clean sheets. We were incredibly grateful to Wayne, who then drove the 300 kilometres back to Sydney because he didn't want to take up a bed that a weary firefighter or fire victim might need.

On New Year's Day 2020 our ten fire trucks and two command four-wheel drives set off early in convoy to return to Sydney, negotiating many police roadblocks and fallen trees that littered the highway. Local council and energy authority crews were already out in force clearing the roads with chainsaws, but it was obvious that reconnecting power would be a massive undertaking with hundreds, maybe thousands, of power poles destroyed.

The devastation was incredible. There was no greenery at all. Fires had blasted through to the ocean on massive fronts, leaving a smouldering wasteland. Everything had been consumed, even the tops of the tallest trees, testament to the ferocity of the crown fires that had crossed the highway as if it wasn't there.

I thought back to my horror the day before (only one day?) when I'd seen stationary lines of traffic trying to head north to safety. I fully expected to see burnt cars with crime scene tape around them, marking the site of fatalities. I was relieved and grateful that there were none. I marvelled at the massive effort that the police must have put in the day before to clear the highway of vehicles, evacuating people as the converging fire fronts raced to the coast driven by gale-force winds and fire-generated storms. To me they were unsung heroes, given the likelihood of a large death toll. Any death is one too many, and every person who dies leaves behind a devastated family and circle of friends. But given the horrendous conditions on New Year's Eve, it had to be conceded that there were relatively few deaths.

I had seen the police in action at Batemans Bay, where they placed themselves in harm's way by parking across key roads as flames roared towards them, and turned around cars so that motorists didn't drive to their deaths. I hoped they would be recognised along with the firefighters from all of the New South Wales agencies, from interstate and from overseas. And along with the paramedics, too. Wherever firefighters go, so too do paramedics, tending to burns; red, smoke-affected eyes; cuts; sprains and broken bones.

A couple of times on the drive home we stopped to deal with small fire outbreaks on the side of the highway. Probably, in hindsight, a pointless exercise, but it made us feel better to be doing something practical among such widespread devastation.

Snippets of news started to come through as we headed further north, stopping for a Macca's breakfast at Nowra where people spontaneously came up and thanked us as we sipped our coffees. The news confirmed what I feared – hundreds of homes lost, a number of confirmed fatalities, and people unaccounted for. This was on top of the hundreds of homes already lost in the north of the state. In Victoria and South Australia, hundreds more had been destroyed. The fires had now far exceeded the worst previous property loss from fires in Australia by a large margin, ultimately about 1,000 more homes destroyed than any previous fire season through history. The scale of losses in the natural world could not be tallied at that stage, but far outweighed the carnage visited upon humans.

Mercifully (I struggle with this word when describing loss of life), the eventual, terrible death toll of thirty-five directly related to fires was less than previous fire disasters such as Victoria's 2009 Black Saturday (173 killed), Ash Wednesday in Victoria and South Australia in 1983 (75 killed), and Black Tuesday in Tasmania in 1967 (62 killed). I put this down to early strategic decisions to order mass evacuations, to effective public messaging and awareness leading up to the worst fire days, and to the use of emergency warnings that sent messages to mobile and landline phones. However, later studies found that at least 417 more deaths could be attributed to the impacts of smoke.

The year 2020 was off to a bleak start. I tried to tell myself that things surely couldn't get any worse. Little did we know at that time that it would prove to be one of the most challenging and unusual years, with fires, storms, floods, a global pandemic, a recession, and later, even a mouse plague.

The fire weather was relentless. January saw more days of Extreme and Catastrophic Fire Danger, more out of control fires, more property loss and more deaths.

Further catastrophic fire weather was forecast for 4 January, and evacuations were ordered in a number of areas. This was the day that the Prime Minister held a press conference, and apparently without reference to state premiers or fire chiefs, announced the call-out of Army Reserves. He also announced $20 million towards four additional large firefighting aircraft, stressing that fire chiefs had only requested two. Ultimately, these aircraft didn't arrive in time to assist firefighting efforts. ELCA members noted with frustration that this money would have been put to very good use by fire services had it arrived when we had first asked for it, back at the start of the season, and even better, when current fire chiefs had requested more funding in 2018.

At the same time, the PM released a slick Liberal Party advertisement praising his government for their handling of the fires. It was quickly withdrawn when reactions from the public were less than positive. Unfortunately, they could not spin their way out of a real disaster: their performance shortcomings were on display for the world to see.

During January, southern suburbs and small towns in the ACT were besieged by a fire accidentally started by a military helicopter in the Namadgi National Park. Other fires in the New South Wales Riverina, including the Dunns Road fire that impacted the towns of Batlow and Tumbarumba, destroyed at least 182 homes.

Throughout the bushfire siege, authorities stated that due to the protracted nature of extreme weather, the number of fires and their sheer size, firefighters would not be able to contain and extinguish the fires until rain, and lots of it, arrived. The prayers started to be answered in early February.

It is hard to describe what it feels like to a firefighter when rains arrive after a long campaign fire season. Relief, gratitude, the sense that a huge weight has been lifted off your shoulders. Plus emptiness and a feeling of disbelief, because you've been geared up for action for so long, and then it all stops abruptly. Going back to your normal life can seem mundane and meaningless, and some people suffer because they feel as though they're not needed or appreciated anymore. And grief. Grief for lives lost, for the homeless, for the animals, for the charred landscape. When you're fighting fires you don't have much time to reflect on this, and, frankly, you don't want to let your guard down to your own emotions, because it can be overwhelming. A lot of emotions surface once the fires are out, and there was an especially big increase in the need for mental health support and interventions at the end of Black Summer.

In the first week of February 2020 there were torrential downpours and storms that caused flash-flooding and loss of power to thousands of homes in some areas. Up to 700 mm of rain fell at some locations, and after three days many of the biggest fires in New South Wales, including the massive Gospers Mountain fire, and the Green Wattle Creek fire south of Sydney, were declared to be out. This seems to have become a pattern too – long, dry periods followed by intense downpours. Just like in Queensland in 2018, when massive bushfires were followed by massive floods near Townsville which killed hundreds of thousands of cattle.

Climate change drives weather extremes, and it is well established that for every degree of temperature increase, the atmosphere can carry 7 per cent more moisture. The problem is that the rain is not uniformly spread, with many areas of Australia suffering long-term reductions in rainfall, interspersed with short, sharp torrential downpours, much of which simply runs off before it can soak into the ground.

As Australia started to lick its wounds and assess the damage and scale of recovery efforts required, the Bureau of Meteorology gave early indications that the rest of 2020 might be quite different, with computer modelling suggesting high likelihood of a La Niña event, normally involving cooler, wetter conditions. This was welcome news, until New South Wales was again hit by flooding in March 2021.

A lesson from the past is that when the arid centre of Australia receives significant rainfall, fire inevitably follows. The 1974–75 fire season saw millions of hectares burnt in every state and territory owing to widespread, heavy rains in 1973 and 1974 that made the red centre of Australia green. Huge parts of Australia were once again green by the middle of 2021; however, they will inevitably turn brown as grasslands cure and die, at some stage ripe for grass-fires. Given the worsening of fire weather, such fires may end up being quite different to the relatively harmless, meandering ones of the mid-70s.

The 2019–20 bushfire season was Australia's worst by far. Among the thirty-five people killed, were six firefighters; 3,094 homes were destroyed; thousands of other buildings such as schools, shops and farm buildings were destroyed; upwards of 24 million hectares were burnt across Queensland, New South Wales, Victoria and South Australia, and billions of native animals were killed. Smoke killed at least 417 people and resulted in more than 4,500 additional hospital presentations during the fires.[42] Economic losses were in the billions of dollars. The proportion of forest burnt was unprecedented.[43]

The word 'unprecedented' has been used so often in relation to these fires that people could be excused for becoming weary of it, but it was entirely accurate because there were so many departures from the norm, namely:

- the very early start of the fire season in NSW and Queensland
- the extreme length of the fire season, from July 2019 to March 2020
- the severity of fire weather despite the absence of the amplifying effect of El Niño
- the record number of days of Very High Fire Danger and above experienced, particularly in spring
- the record number of days of Severe, Extreme and Catastrophic Fire Danger
- the number of Total Fire Bans declared
- the record number of previously rare fire-generated (pyro-convective) storms
- the sheer size of some of the fires, including Australia's largest recorded forest fire, Gospers Mountain
- the number of human lives lost through fire and smoke impacts
- the number of native animals killed
- the number of homes and other buildings destroyed
- the mental health impacts on firefighters and communities
- the proportion of eastern broadleaf forest burnt
- the amount of rainforest burnt, previously considered immune from intense fires
- the smoke impacts on major cities and regional centres
- the economic impacts, such as the estimated $4.5 billion lost in New South Wales tourism revenue alone.

Australia had never experienced anything like it before, but, unfortunately, expert inquiries and scientists have concluded that the extreme weather that drove Black Summer is a taste of what we can now expect during our periodic worst fire seasons, due entirely to relentless warming.

Ominously, all of the independent inquiries concluded that we will experience even worse fire seasons in future as the planet gets even hotter. There is no question scientifically that the warming is driven by continuing greenhouse gas emissions from human activities, and yet there is a continuing policy vacuum in Australia to drive emissions down sufficiently or to join the rest of the world in trying to stabilise warming.

Notably and thankfully, although the 2019–20 bushfire season followed a serious drought, it did not coincide with an El Niño event which would almost certainly have made it even worse. While the fire season was frightening, it is chilling to contemplate a similar scenario with a strong El Niño in the mix. Something that is inevitable in the future.

9

THE AFTERMATH

AS THE SEASON WENT from bad to worse, and relentless, record-breaking fire weather assailed every state and territory, political rhetoric started to change on the subject of climate change. After mocking dismissal and offhanded rejection, the government evidently started to see that the general public was realising, as the disaster unfolded, that nothing like these fires had ever happened before. An Australia Institute study after Black Summer found that 82 per cent of Australians believed that climate change was dangerous and a factor in the fire disaster following Australia recording its hottest, driest year on record, and after so many people saw the impact of out-of-control climate change unfold in front of their eyes.[44]

The more optimistic of us within ELCA hoped that the tide of public outrage would result in a reasoned and meaningful change of policy in Australia to start to mitigate the threat of climate-change-driven bushfires. With the federal government seeming more interested in polls than policy, I was hopeful that changes in public understanding would drive political change. Instead, the vested interests who benefit in the short term from Australia's appalling climate policies went on the attack.

Throughout Black Summer, a lot of incorrect information was circulated on social media, by sections of traditional media, and by some politicians. There is no doubt that much of this was well organised and intentional, aimed at taking attention away from what was actually happening (worsening extreme weather, driven by climate change, making bushfires uncontrollable), and who was actually to blame for a lack of action that could reduce the risk in future. Such as the federal government, who held back funding for firefighting aircraft, and whose so-called emissions policies have no emissions reduction targets beyond 2030, and even then, are among the world's weakest.

It's vital to understand and debunk the most serious bushfire and climate myths that seek to hide the truth and deflect attention from efforts to protect the land and waters and all the living organisms under threat, which includes us and future generations of Australians.

Myth 1: 'Greenies stopped hazard reduction burning'

One of the more common assertions made during the fires was that backburning (sic; use of this incorrect term is an immediate indicator that those using it have no understanding or direct knowledge of bushfires) was stopped by greenies. During the fires, Nationals MP Barnaby Joyce was particularly vocal, claiming that 'greens policy' gets in the way 'of many of the practicalities of fighting a fire and managing it'.[45]

Understandably, such claims sparked outrage. The public felt powerless against the forces of Mother Nature and great sympathy for those who were suffering. Identifying a culprit enabled people to focus their energies and raw emotions on something tangible. However, the claims have no basis in truth:

1. The Australian Greens is not and has not been in power in any state or territory except the ACT and remains a minority party. It does not control government policies in Australia and has never had that ability.

2. All laws, policies and frameworks for hazard reduction burning have been developed and are administered by Labor and Liberal / National Party governments in the states and territories, with the exception of the ACT where the Greens are in coalition government with Labor. For this myth to be true, Greens would have had to somehow take the reins of power from Liberal / National Party coalition and Labor governments, or somehow force them to support legislation and regulations restricting hazard reduction. This has never happened and is unlikely to ever happen in the future.

3. Many conservation (i.e. 'green') organisations, for example the Nature Conservation Council of NSW, are not anti-burning. To the contrary, they have scientifically valid pro-burning policies recognising that many Australian plant species require regular fire to survive and also that ecological requirements must be balanced with the need to protect human life and property.

4. The NSW Bushfire Inquiry commissioned into the 2019–20 season found that, on average, fuel levels were no different than they had been for the last thirty years.

5. All independent inquiries found that the 2019–20 fires were propelled by unprecedented fire weather, not by excessive fuel loads.

6. Tens of millions of hectares were burnt across Australia; 5.4 million hectares in New South Wales alone. Under current resourcing and funding arrangements it would be impossible to conduct enough hazard reduction to materially affect fires of that size and extent.

7. Agencies such as the National Parks and Wildlife Service produced records showing that they had on average been conducting more hazard reduction burning in recent years than in the past.

8. The Royal Commission, NSW Bushfire Inquiry and other investigations explained how bushfires burning under Extreme and Catastrophic Fire Danger conditions are not slowed down or controlled on reaching land that has been subjected to hazard reduction. It has been known for many years that hazard reduction is vital and must be part of the mitigation approach, but it is just one, of course very important, piece of a larger puzzle because Extreme and Catastrophic Fire Weather conditions that reduce the mitigating effects are becoming more common.

9. Climate change is reducing opportunities to safely conduct prescribed burns in some locations where the warming and drying trend has reduced the length of time available each year to safely conduct burns. When it is not too dry and windy to burn, it is often too wet for fires to start, because for each degree of warming, the atmosphere can hold 7 per cent more water vapour, often resulting in heavier rainfall over shorter periods. Longer fire seasons, up to three months longer in some parts of Australia, are steadily reducing windows of opportunity to burn.

10. A recent research paper suggests that new windows of opportunity for burning in south-east Australia could open in August and September from about 2060, but confirms that traditional east coast burn times, March to May, will become increasingly difficult for burning due to climate change.[46] The possibility of better times for burning in 40 years is of little comfort and of no practical and immediate use.

11. Some states rely substantially on volunteer firefighters to carry out hazard reduction burning, and in most states permanent workforces formerly employed by government forestry organisations have reduced. This can mean that a lot of burning is restricted to weekends when the largest number of volunteers are available, further limiting time available for burning.

Myth 2: 'We've had weather conditions like this before'

No. We haven't. The 2019–20 fire season was driven by hot, dry, windy weather conditions, sustained for months. Many weather and fire danger records were broken by large margins, and extreme fires burned for months longer than previous fire seasons, with 2019 being the hottest, driest year ever recorded in Australia.[47]

The Bureau of Meteorology occasionally publishes detailed summaries of significant weather and climate events. Their *Special Climate Statement 72 – dangerous bushfire weather in spring 2019* detailed just how extreme the weather had been during spring 2019. It found that, 'Across Australia as a whole, spring 2019 saw the highest fire weather danger as measured by the Forest Fire Danger Index with record high values observed in areas of all States and Territories, while rainfall in spring 2019 was below average to lowest on record over many areas of Australia – 62 per cent below average across the country as a whole.'[48]

Extreme heat and fire weather across the new-year period persisted into summer. In December 2019 there were eleven days in which the national area-averaged maximum was 40°C or above. Prior to December 2019 there had been only eleven such days recorded since 1910, seven of these occurring in the summer of 2018–19. Think about that for a moment: in just one disastrous summer, we recorded the same number of extreme heat days as had been experienced over the previous *109 years*.

Myth 3: 'The fires were started by arsonists and had nothing to do with climate change'

A number of conservative politicians strongly asserted during the fires that intentional fire-setting by arsonists was a major issue, or that arson was more of a problem than climate change.

Nationals MP George Christensen posted on Facebook:

> *Those politicking off of (sic) the bushfires disgust me. The rot being spewed by the Left and many in the media is wrong and it's disrespectful to those who've lost everything. [. . .] The upshot is that the cause of the fires is certainly man-made, it's just not man-made climate change. It's man-made arson that, to me, almost borders on terrorism. And it's man-made stupidity by politicians who support laws that stop backburning [sic] and decent fire breaks.*[49]

The strident claims coming from people whom many might assume to be knowledgeable sources inspired, as could be expected, calls for immediate action and jailing of the perpetrators. The claims were reinforced by a torrent of social media posts from sometimes dubious sources.

Research by Dr Timothy Graham, senior lecturer on social network analysis at the Queensland University of Technology, identified strong indications of a deliberate campaign of disinformation by unidentified actors during the fires. His research found that popular hashtags used to track the fires on social media attracted a suspiciously high number of 'bot-like and troll-like accounts', deliberately undermining the link between the current bushfires and the longer, more intense fire seasons brought about by climate change.[50]

Basic research readily established that the assertions, just like untrue claims about fuel loads and hazard reduction burning, were baseless.

The NSW Bushfire Inquiry identified that during the bushfire season there were 11,774 bushfires in New South Wales. NSW police reported that there were just eleven instances of arson. The Inquiry also listed thirty-two of the largest, most damaging fires that occurred in New South Wales, together with their determined causes. None of the major fires were attributed to arson. The vast majority (twenty-four) were started by lightning, with the rest started by factors including 'powerlines', 'shredded tyre', 'equipment', and 'debris burning'.[51]

Chapter 2 of the Inquiry Report looked in detail at the causes of fires. It found that while arson was suspected in some instances, those fires accounted for a very small proportion of the total area burnt. Police Strike Force Tronto found that of fifty-nine fires found to be deliberately lit, only eleven were started with the intent to cause a bushfire.[52]

The RMIT ABC Fact Check Unit found that offences for bushfire arson during the fires were below long-term averages in New South Wales and Victoria, with Victoria recording the 'lowest level for at least a decade, and well below the ten-year average of 49.5 offences'.[53]

A notable feature of these fires, many of which grew to be very large, such as the Bees Nest fire that scorched 114,000 hectares through September, October and into November 2019, was that they were mainly started by dry lightning storms. This reinforced research findings that higher temperatures, low humidity and drier fuels make ignition by lightning more likely.[54]

Myth 4: 'Grazing in National Parks will reduce fuel loads'

This is another recurring claim that arises during major bushfire seasons, and one that has been championed in New South Wales by the National Party.

Scientific studies have found the claims to be baseless and without merit and that grazing in fragile ecosystems still struggling to recover from the massive 2003 alpine fires is potentially dangerous and damaging. Added to this is the level of damage from feral horses that are multiplying dangerously due to policies of the NSW Government, strongly promoted by the National Party.

The Victorian Alpine Grazing Taskforce in 2005 found that, 'the scientific research is adequate and consistently reveals that grazing has a deleterious effect on biodiversity',[55] leading to a ban on grazing in Victoria's alpine parks. The report explained that, 'the most flammable fuel types in the park, which contribute virtually the entire available fuel load to wildfires, are branches, twigs, bark, eucalyptus leaves and shrubs. With the exception of some shrubs, *cattle do not eat these fuels*,' (emphasis added) and further, 'It was also pointed out that cattle eat the new green shoots and not the dead, dry grass that constitutes the more flammable component of the fine fuel,' concluding that, 'cattle grazing does not make an effective contribution to fuel reduction and wildfire behaviour in the Alpine National Park.'[56] The same is true in other forested areas, not just in national parks.

Myth 5: 'We've had worse fires before'

One of the key pieces of disinformation aired after Black Summer was that we've had worse fire seasons and bushfires in the past. Specifically, that the 1974–5 fire season was much worse.

There can be no valid comparison made between the extensive nationwide grass fires in 1974–5, and forest fires in 2019–20. While sometimes faster-moving, grass fires are less intense and less damaging than forest fires.

The 1974–5 fire season across Australia was unique, because widespread, unusually heavy rains through 1973–4 transformed the arid centre of Australia into lush grasslands. A dry spring in 1974 dried out the grass so that by the time summer arrived it was tinder dry and ready to burn.

The first fires started in the Barkly Tableland area of the Northern Territory in June and July and burned around 2.4 million hectares. Fires then broke out in every state and burned for months, with an estimated 117 million hectares ultimately burnt across Australia. Very few buildings were damaged and stock losses were surprisingly light, probably because the fires were not driven by strong winds. Most fires were simply left to burn, with no suppression action taken due to their remoteness and low risk.[57]

Other claims were made on social media that the Victorian fires of 1851 were far worse than those of Black Summer. This claim is irrelevant and baseless. There are no reliable weather records from that time (BoM temperature records started in 1910), and very little information available about fire impacts other than what can be inferred from limited news reports of the day from the new colony. There were no central agencies able to verify or tally losses, and Australia was still decades away from Federation. It is easy to make claims like this because they can neither be confirmed nor denied.

Clearly, there were widespread fires throughout Victoria in 1851 and many settlers were burnt out. Conditions were very different back then. Victoria was a sparsely populated new colony, and there were no firefighting or forestry organisations able to respond. Fires burned wherever the wind and weather took them until rain fell, and settlers had to rely on their own best efforts for protection.

Fire experts and historians know little about the 1851 bushfires, certainly not enough to validly compare them with recent events. People making such claims on social media and other platforms know even less. Their assertions should be ignored.

*

The fact is, each of the major inquiries into the causes of the 2018 and 2019–20 fire seasons found that climate change was a driving factor. They also found that we can expect similar conditions – or worse – in future, because of escalating human-caused warming.

What the experts said – an even more dangerous future

Our premier national scientific bodies, the Commonwealth Scientific and Industrial Research Organisation (CSIRO) and Bureau of Meteorology (BoM), have described the processes that will lead to ever-increasing warming and therefore an increase in serious fire weather over the next two to three decades, regardless of what we do about greenhouse gas emissions. Further warming is locked in over that timescale due to greenhouse gases already emitted. The Bushfires Royal Commission stated that if we fail to take strong action on emissions now, conditions will then continue to worsen rather than stabilising and eventually improving. This is why immediate strong action on emissions is crucial.

We are already experiencing longer fire seasons, smaller windows in which to conduct fuel-reduction activities, more days each year of serious fire weather, a reduction in the number of mild fire seasons, and major increases in the frequency and severity of serious fire seasons. The South Australian bushfire inquiry stressed that the fires were so intense they 'exceeded the limits of firefighting capacity', also describing in scientific detail why fires can now burn intensely overnight, as if it is daytime.[58] This in itself is a major change in previously well understood fire weather and fire behaviour conditions, reducing the effectiveness of firefighting tactics targeted at milder night-time conditions, also increasing the risks of backburning.

Catastrophic bushfire seasons like Black Summer will become more and more likely as the planet continues to warm, with one study concluding that the record weather conditions of 2019 will be 'average' by 2040 and 'exceptionally cool' by 2060 under current emissions trajectories.[59] The Victorian bushfire inquiry found that the conditions experienced during Black Summer were 30 per cent more likely than a century ago, due to warming, and that the likelihood of such conditions in future will rise four-fold if global temperatures increase by 2°C or more. Australia has already warmed by 1.4°C since 1910.

The NSW Bushfire Inquiry established that we could easily see even worse bushfire conditions in future. A strong El Niño event, which was absent in 2019–20, and fires breaking out near large urban areas untouched by fires during Black Summer – Sydney, Newcastle, Wollongong and the Central Coast – could result in even greater loss of life and property. Return of an intense El Niño is unfortunately inevitable because these events are set to become more frequent and intense as the climate warms.[60]

Up until 2019–20, the worst Australian bushfires in terms of death and destruction had taken place in Victoria and Tasmania, not in New South Wales which had never before lost thousands of homes in a single fire season, unlike Victoria in 1898, 1939, 1983 and 2009, and Tasmania in 1967.

With New South Wales experiencing property losses during Black Summer over 1,000 per cent greater than the worst previous fire seasons through history, it is chilling to imagine such an escalation in losses in Victoria and Tasmania, but unfortunately it is not beyond the realms of possibility. After all, California's loss of 20,000 buildings to bushfires in 2018 after losing 10,000 the year before then another 10,000 in 2020 is a chilling foretaste of what might happen here as conditions worsen and become even hotter.

Settled science – how climate change caused Black Summer

Atmospheric warming was first proposed by French mathematician and physicist Joseph Fourier in 1824, and in 1859 Irish physicist John Tyndall demonstrated the ability of water vapour and carbon dioxide (CO_2) to absorb and store radiant heat from the sun as it entered Earth's atmosphere. In 1896 in Sweden, Svante Arrhenius quantified the warming effect of CO_2 and made predictions of global warming based on a doubling of CO_2 in the atmosphere. The term 'greenhouse effect' was first used in 1901. The concept is not new and the basic science underpinning it is not disputed, despite a relentless campaign of misinformation.

About 26 per cent of radiant heat from the sun that reaches Earth is reflected by clouds and the atmosphere back into space, with about 19 per cent absorbed by the atmosphere. The remaining heat is absorbed by the Earth's surface, particularly our oceans. Greenhouse gases absorb heat that would normally radiate back into space from the surface. Without this effect, the Earth would be very cold and would not support life as we know it.

Levels of CO_2 in the atmosphere were very stable for millions of years, up until the start of the Industrial Revolution which brought about the widespread introduction of mechanisation using steam engines fuelled by the burning of wood and coal. Climate scientists use a baseline of the mid 1800s as a comparator for levels of CO_2 in the atmosphere. The burning of fossil fuels such as coal, oil and gas produces large amounts of CO_2. As mechanisation spread across Great Britain, Europe and the United States of America, so too did the burning of coal, then the burning of oil and gas. The Industrial Revolution brought great prosperity and innovation and was accompanied by significant increases in population, exponentially increasing demands for energy and therefore of emissions, as humans burned more and more fossil fuels.

As the burning of these fuels increased across the globe, hastened by invention of the internal combustion engine and the discovery and transmission of electricity mainly generated from the burning of coal, levels of CO_2 in the atmosphere rapidly increased.

As had been demonstrated back in the mid-1800s, increased levels of CO_2 and other greenhouse gases in the atmosphere inevitably resulted in heat being captured, resulting in gradual warming. Methane and nitrous oxide absorb more heat per molecule than CO_2, and increasing extraction of commercial gas as well as the thawing of permafrost in both the Antarctic and Arctic, releases significant amounts of methane and soil carbon that in turn contribute even more to warming.[61]

The level of CO_2 in the atmosphere back in the mid-1800s was about 280 parts per million (ppm). According to the US Government's National Oceanographic and Atmospheric Administration (NOAA), carbon dioxide levels in the atmosphere reached a milestone 421 parts per million in April 2021, the highest concentration in 3.6 million years. NOAA says that the current annual rate of increase in CO_2 levels is now 100 times higher than previous natural levels, for example during the ending of the last Ice Age around 11–17,000 years ago when natural emissions were quite high.[62]

Claims that there are other explanations for rapidly increasing levels of greenhouse gases in the atmosphere and for rapidly increasing global temperatures have no scientific basis. Such claims seem to be aimed at delaying action on emissions that would harm profits of fossil fuel producers. It has been firmly established and demonstrated that climate change is being caused primarily by humans through the burning of coal, oil and gas, and as a result we are experiencing an accelerating global catastrophe of our own making. There is worldwide scientific agreement that warming is human-caused, anthropogenic, hence the term Anthropocene, to describe the most recent period of Earth's history when human activity started to have a significant impact on the planet's climate and ecosystems. Claims of scientific disagreement are fabricated and dangerous.

More than 90 per cent of the excess heat in the climate system has been absorbed by the oceans.[63] Alarmingly, the rate of ocean warming is accelerating and it seems that oceans are becoming less able to soak up excess CO_2, as large-scale worldwide clearing of forests destroys trees that absorb CO_2 from the atmosphere, reducing the capacity of another significant carbon sink. Australia has a sorry record of large-scale clearing of forests, and this has accelerated in recent decades.

Climate change is increasing bushfire risk

There is an undeniable and proven link between climate change, extreme weather and bushfires. The CSIRO and BoM have established that Australia is warming slightly faster than other parts of the world, so we are on the front line of climate-driven disasters.[64] The warmer background climate means that we are experiencing more extremes, as happened in 2019 – the hottest, driest year ever recorded in Australia.[65] Australia is one of the countries most vulnerable to climate change in the developed world.

Worsening fires are not just a problem here. As I witnessed during my observations of bushfire conditions and activity in the USA, Canada, Spain and France – supplemented by studies at the US Fire Academy and roles of Fellow of the UK-based Institution of Fire Engineers and Director of the International Fire Chiefs' Association of Asia – human-caused climate change is a crisis on a global scale.

Australians are recognised internationally as being smart and resourceful. We have some of the best firefighters, best firefighting arrangements and best fire scientists in the world to guide us. However, I fear that despite our best efforts from now on, our response and recovery efforts will increasingly prove to be almost futile in the face of the worst that Mother Nature will increasingly throw at us in a rapidly warming, super-charged climate. Black Summer does not yet represent a new normal, but it clearly represents a new extreme. Projections of temperature increases point to fires like this representing not an extreme, but an average year before the middle of this century.

California is another canary in the coal mine, and possibly also the best-resourced place on the planet in terms of firefighting equipment and number of firefighters. When I first visited in 1995 and deployed to fires around Los Angeles and in San Diego County, they considered the loss of about 4,000 structures during a fire season to be catastrophic. Just prior to our own 1994 bushfire emergency the state endured the 1993 Southern California firestorm.

In 2017, Californian fires destroyed nearly 10,000 buildings, then in 2018, 20,000 buildings were lost with almost 100 people killed. In August 2020, Northern California was besieged by unprecedented dry lightning storms, with 14,000 lightning strikes in one day starting dozens of fires. The storms coincided with an unprecedented heatwave and strong winds.

For the first time in US history, on 17 August 2020, a fire tornado warning was issued as a large fire generated its own weather system and the massive convection column started to become an enormous twister. US media referred to Canberra in January 2003 where the first recorded large-scale fire tornado occurred.

In September 2020, after thousands of structures had already been destroyed and twenty-three major fires burned out of control, record temperatures of almost 50°C were experienced in Los Angeles, and many other urban centres had the highest temperatures ever recorded. September 2020 was the hottest on record in California and around the world. In fact, we haven't experienced global temperatures at or below average since 1977. Each decade is warmer than the previous one, and the 10 warmest years have occurred since 2005. Temperature records continue to tumble as atmospheric concentrations of CO_2 march on unabated.

During August and September, fires continued to break out and grow in US states including Oregon, Idaho, Colorado, Washington State and California. Early in September, California surpassed the largest area ever burned in a single fire season: nearly two million hectares, or more than two thirds the size of Hawaii. This was incredibly significant, given that historically California's worst fires and greatest losses of life and property tended to happen in October when the Santa Ana winds blew across Southern California. It was of enormous concern that nearly 9,000 structures and thirty-one lives had already been lost by the end of September, with no reprieve in sight.

The US is not the only country experiencing worsening wildfire problems driven by rapid warming. Countries accustomed to periodic fires such as Canada, Chile, Greece, France, Spain and Portugal are all experiencing worsening conditions with increasing losses of life and property.

At the same time, countries that had not previously been considered fire-prone have developed bushfire problems. In 2018, Sweden had to call for international help to fight out of control forest fires on a scale never faced before. In 2020, for the second year in a row, major fires raged in the Arctic Circle in Siberia where temperatures over 40°C were experienced. In the tropics across Malaysia and Indonesia and in the Amazon, major fires are now commonplace. Even cold and wet countries like the United Kingdom and Greenland are starting to see regular landscape fires.

A recent study found that the UK is experiencing more wildfires as a result of climate change, and that projected warming could see the number of fires increase by a factor of four by 2080.[66] I have travelled to the UK on three occasions since 2004 to speak to fire chiefs there about bushfire control, as they come to grips with what previously had been a rare hazard.

This highlights how weather extremes are worsening globally, with more widespread fires of greater intensity being one of the symptoms. Dry periods are drier with increasing rates of evaporation due to higher temperatures, heatwaves are hotter and more frequent, winds are stronger, and storms are more damaging.

US fire historian Stephen Pyne coined the term Pyrocene, or era of fire. In a discussion following Black Summer he explained to me how human use of fire has shaped the world over hundreds of thousands of years. The natural world is now paying a heavy price due to the unprecedented combustion of ancient fossil fuels that were deposited over millions of years. Humans have burned massive amounts of these fuels over just two centuries – and natural systems have no capacity to deal with the output. Where this will lead is unclear, but at this point in the journey it is safe to say that it will not be to a good place.

The dumping of vast quantities of CO_2 into the atmosphere reminds me of the first environmental disaster I encountered in my childhood. Back then, more than a century of dumping raw sewage into the ocean had turned the formerly pristine Northern Beaches of Sydney into murky health hazards.

The colloquial terminology used to describe floating turds that often appeared beside you in the surf was 'blind mullet'. Skin rashes and ear infections were common. Eventually the government was forced to concede that the capacity of our oceans to absorb waste was finite, and that proper treatment of effluent was necessary in order to protect people and the environment. As is the pattern in our political life, the effects of our negligence and lack of care for other species was ignored until a majority of ordinary citizens were being directly affected.

Our atmosphere is now also full of our crap. We are starting to notice, but the consequences are far worse than having to share waves with the odd blind mullet. Since CO_2 is colourless, odourless, tasteless and invisible, these qualities may have helped vested interests to deny the dangers of emissions and deny that climate change is real. Hopefully the tragedy of Black Summer and what we all lived through will start to change that.

Parts of Sydney reached 50°C during Black Summer, something that I thought I would never see. Back in the 1970s, bushfire experts like Alan McArthur (who developed the Forest Fire Danger Index (FFDI)) established that, theoretically, the worst fire weather likely to be experienced in Australia at that time and lead to the maximum FFDI value of 100, would involve a temperature of 40°C.[67] We now regularly exceed an FFDI of 100, and after 2009 had to introduce an additional fire danger rating representing 'off the scale': Catastrophic. The implications of increased temperatures on fire-fighting can be profound. As a young firefighter I was accustomed to dealing with fires driven by temperatures in the mid- to high 30s, perhaps even into the low 40s. When low humidity and high winds are coupled with temperatures of nearly 50°C, however, bushfires are an entirely different proposition to what we had to deal with in a cooler, more benign climate.

As Australia becomes hotter and drier, forest and grassland fuels are drying out and becoming more flammable. Higher year-round temperatures increase evaporation, exacerbating the intensity of droughts and further drying out soils and vegetation. Fire seasons are starting earlier, finishing later, and the average number of days each year of serious fire weather continues to increase, with temperature a major factor. Hot days bring fuels closer to their ignition temperature, meaning that less energy is needed to make them burn. In turn, rates of fire development and speed of fire spread increase because adjoining drier fuels ignite and burn more readily.

Average temperature increases over time can be deceptive, as they mask temperature extremes. Overall warming of the atmosphere raises the baseline, meaning that extremes become more likely and that when they occur they can be higher than previously experienced. Since 1950 the annual number of record hot days in Australia has doubled. Heatwaves are becoming hotter, lasting longer and occurring more often.

The best analogy is a dam: if water levels in a dam are low, it takes a lot of rain to make it overflow and cause a flood downstream. If the dam is already quite full, it takes far less rain to cause an overflow. The baseline is therefore vitally important. Baseline temperatures have been rising implacably, accelerating ever faster from 1960.[68]

A long-term drying trend across south-east and south-west Australia has dried out the landscape, making bushland more flammable and easier to ignite while increasing amounts of available fuel. This has been most evident in the south-west and south-east of the country, where rainfall analysis extends back to around 1890, allowing the comparison of modern droughts to historical dry periods.[69]

In 2019, rainfall from January to October was 70 to 80 per cent below average in some locations in northern New South Wales.[70] These were the areas where fires started to burn from July and August 2019, four months before the start of the official New South Wales fire danger season and where major fires had burned in February and March 2019.

The dry conditions throughout the year followed prolonged reductions in rainfall across most of south-eastern Australia since the beginning of 2017 during the worst drought on record. The higher temperatures experienced over that period resulted in increased evaporation, amplifying the effects of the drought and making soils critically dry.[71] There were many observations throughout 2019 of native forests appearing to die off and foliage going brown in areas as diverse as south-east Queensland, northern New South Wales, the Southern Highlands, the far south coast of New South Wales, and East Gippsland in Victoria.[72]

Rainfall in Victoria in 2019 was about 28 per cent below average, and most of Gippsland, which was severely impacted by fire, had annual rainfall totals in the driest 10 per cent of records.[73] The 2017–20 drought, though shorter in duration than other serious droughts, broke records in terms of intensity.

Drought conditions, long-term reductions in rainfall and higher rates of evaporation as a result of higher year-round temperatures can all increase the amount of available fuel because dead vegetation has lower moisture content and even some standing, living fuels will dry out and be available to burn. The more fuel available to burn, the higher the fire intensity and heat output, therefore the stronger the convection column that carries embers aloft and can reach into the stratosphere, sparking fire-generated storms. Drought can also cause a significant increase in ground fuel levels as stressed trees shed bark, leaves and small branches. Even a small increase in the amount of available fuel can have a significant impact on fire intensity, rate of spread, and the ability to generate embers and spot fires.

During the 2019–20 season there were many reports of fires burning at intensities never seen before, and satellite imagery confirmed that the radiative power, a measure of heat output in kilowatts per metre of flame front, was the highest ever recorded for every day of spring and summer during 2019–20.[74]

In the early 2000s, amid rising temperatures, US scientists started to talk about heatwave droughts, where the rapid onset of drought conditions influenced by high temperatures and high levels of solar radiation quickly led to low soil moisture content. These are now termed flash droughts.[75]

I have seen this rapid drying and warming process myself in New South Wales preceding the January 1994 statewide bushfires, and also the hot, dry weather preceding the New South Wales fires in October 2013. In our new climate, it no longer takes a sustained drought for an area to become dangerously dry. Drought conditions can now set in, from a meteorological perspective, in the blink of an eye.

I described how, as a firefighter, I observed that serious fire seasons were happening more frequently. Queensland is a good case study of a state that has seen a rapid worsening of bushfire conditions associated with increased frequency of major bushfire seasons. Weather conditions have become hotter and drier, fire seasons have lengthened, and more days of Very High Fire Danger and above are being experienced than in the past.[76]

The Blue Mountains region west of Sydney is acknowledged by the Insurance Council of Australia as one of the most dangerous and exposed in Australia in terms of property risk from bushfires, with hundreds of homes situated on ridge tops above steep wooded slopes.[77] The region has a significant fire history with large fires resulting in property loss happening roughly a decade apart throughout the twentieth century. Records show that there were major fires with property loss in 1936, 1944, 1957, 1968, 1977 and 1994. Since then, the frequency of major fires in the

Blue Mountains has changed significantly, reflecting changing climate and weather patterns and a shift towards hotter, drier, more dangerous fire weather. Since the Blue Mountains fires in 1994, major fires have broken out there in 2001, 2002, 2006, 2013, and 2019–20.[78, 79] This represents almost a doubling in the frequency of serious fire weather seasons in the Blue Mountains. Similar shifts are being seen across the nation in many areas.

Tasmania is probably the starkest example where major fire seasons had in the past been relatively rare. Over the last two decades Tasmania has experienced multiple serious fire seasons and fires in areas where it had formerly been too wet to burn.

Bushfire seasons globally are starting earlier and lasting longer as warm seasons increase in length. The Bushfires Royal Commission noted that Australian bushfire seasons have increased in duration by up to three months in some locations, impacting on the ability to carry out hazard reduction burns, the principal mitigation tool available to fire and land management agencies. The official bushfire season in New South Wales, as enshrined in the Bush Fires Act 1949 and then the Rural Fires Act 1997, runs from 1 October to 31 March the following year. For practical purposes, though, bushfire danger seasons now commence in August most years and sometimes extend past 31 March. Longer fire seasons have been accompanied by record-setting periods of serious bushfire danger earlier than previously experienced, such as 6 September 2019, when Extreme and Catastrophic Fire Danger was experienced at multiple locations in New South Wales and Queensland, and April 2018, when a major bushfire burned through southern Sydney suburbs.

Rather than waiting for the legislated start to the bushfire danger period on 1 October, twenty-one local government areas in New South Wales commenced their Bushfire Danger Period in

August 2019, and a further fifty-three local government areas commenced their Bushfire Danger Period in September, because serious fire weather conditions started much earlier than expected.[80] It's a measure that I've seen adopted more and more during the past decade, in reaction to the reality of the changing conditions in which we live.

During Black Summer, major fires burned across Queensland and northern New South Wales from July 2019, which of course was still winter. Fires burned out of control in the Northern Territory and Western Australia in September, and across Tasmania and South Australia in late October. By late November, every state was alight when fires broke out in Victoria following lightning strikes in East Gippsland.

Increasingly longer periods of serious fire weather impact on the ability of fire services to control major fires, because their resources become stretched, limiting the ability to share resources across borders. Serious fatigue among firefighters battling longer and more intense fires is on the rise, and longer fire seasons make it harder for volunteers to be away from their primary employment. The gap is rarely adequately filled by the much smaller numbers of full-time firefighters employed by urban fire and rescue services, national parks and forestry agencies.

A consequence of longer fire seasons in each state and territory is that seasons are increasingly overlapping. As we have seen, Australian fire seasons now also overlap with those in parts of the northern hemisphere, with CAL FIRE reporting that fire seasons in the Sierra Nevada Mountains have lengthened by at least seventy-five days per year.

Over the previous century it was established that Australian fire seasons started in the north and moved south, consecutively affecting different jurisdictions.[81] Fire weather would start in the

Northern Territory during the dry season and then move progressively south, affecting Queensland, NSW, then the southern states and south-western parts of Western Australia. As the fire season ended in one jurisdiction, it would start in the next one to the south.

The sequential nature of the onset of serious fire weather each year enabled fire services to support each other across borders, a cornerstone of Australian firefighting doctrine since the early 1990s. The ability to call upon assistance from other states and territories has been an efficient and effective force multiplier.

Unfortunately, those days appear to be numbered. Simultaneous major fires in different jurisdictions are becoming increasingly common. Overlapping fire seasons reduce the ability of fire services to defend life and property and contain major fires, because they limit the sharing of vital personnel and equipment. The basic issue facing all fire services now is that jurisdictions are becoming increasingly reluctant to release and share resources to assist others when their own state or territory is burning or could be at risk – reducing overall national firefighting capacity when it is needed most.

The South Australian bushfire inquiry noted that its own firefighting resources became stretched because it was assisting in other states, and questioned the advisability of doing this in future. When desperately needed, the other states were not in a position to reciprocate by sending large-scale help to South Australia.

By increasing the frequency and / or severity of destructive weather events as well as background conditions such as average temperature and sea level, climate change is increasing the risk of what are called compound events: instances where multiple destructive events or elements occur at the same time or in close succession, exacerbating one another, resulting in the overall impact being worse than if each had occurred in isolation. Compound events pose a significant threat to the capacity of emergency services to respond effectively (e.g. large fires in southern states and flooding events or cyclones in northern Australia) occurring at the same time or back-to-back.

There has been a measurable increase in the annual accumulated Forest Fire Danger Index (FFDI) across large parts of Australia, meaning that there has been a significant increase in the number of days of Very High Fire Danger and above.[82] BoM fire weather records show that since 1950, when reliable records began, the average number of days of Very High Fire Danger and above in New South Wales during spring is two. The previous record in the spring of 2002 saw eleven days of Very High Fire Danger and above. That long-standing record was smashed, almost doubled, in 2019 with twenty-one days, a number of these being in the Severe, Extreme, and even Catastrophic categories.[83]

During Black Summer in New South Wales there were six days of Catastrophic Fire Danger, two of them affecting Sydney which had never experienced a Catastrophic Fire Danger rating before. There were twenty-two days of Extreme Fire Danger and seventy-two of Severe. These numbers are many magnitudes higher than any previous year in history. There were fifty-nine Total Fire Ban declarations, eleven statewide total fire bans and forty-four bushfire emergencies declared under section 44 of the Rural Fires Act.[84] The annual accumulated McArthur Forest Fire Danger Index (FFDI) in 2019 was the highest on record since measurements started in 1950.[85]

Pyroconvective events (fire-generated lightning storms) were previously quite rare. The Royal Commission found that there had been a total of sixty events in Australia from 1978 to 2018. During Black Summer there were an astounding thirty-five recorded. This represents a deeply concerning increase in the frequency of these formerly rare events. It is also an indication of a significant increase in the intensity and heat output of fires, because pyroconvective storms cannot form without high levels of heat that drive super-heated smoke plumes into the stratosphere.

When fires couple with the upper atmosphere in this way, explosive thunderstorms can form, generating strong down-draughts, tornados, strong wind gusts from multiple directions, and lightning that can start new fires long distances from the main fire.

Veteran firefighters have limited experience of such conditions and current firefighting doctrine is unable to adequately deal with the dangers and variables associated with pyroconvective activity. We therefore have to radically rethink how we tackle the symptoms of Australian bushfires in the Pyrocene, while simultaneously dealing with the cause – greenhouse gases that drive our changing climate.

10

WE MUST STOP THE CLIMATE EMERGENCY BECOMING A CLIMATE DISASTER

BLACK SUMMER, UNPRECEDENTED AS it was, represents what our worst bushfire years will increasingly look like as the world continues to warm. Research has established that massive weather-driven conflagrations will become increasingly frequent, with fires on the scale of Black Summer increasingly common by the middle of the century. The worst future bushfire years will be inconceivable, magnitudes worse than what we experienced in 2019–20. This is frightening to contemplate, but because of the worldwide failure to curb greenhouse gas emissions we are locked into future warming and escalating impacts and have little hope of avoiding such fires. We will have to try to adapt to the consequences, while taking drastic actions to curb emissions if we are to protect future generations and the natural environment from continued escalation.

A recent report by the Australian Academy of Science, authored by our most eminent climate scientists, sends yet another stark warning from the scientific community. 'Current international commitments to greenhouse gas emission reduction, if unchanged, would result in average global surface temperatures that are 3°C above the pre-industrial period in the lifetimes of our children and grandchildren.'[86] Such warming would have profound consequences for Australia and the rest of the world.

In order to adapt, a range of measures need to be taken to try to protect life, property and the environment from escalating risks across the short, medium and long terms.

Short term

In the short term, we need to take actions that will help to reduce risks to exposed communities and increase community resilience, while at the same time implementing measures to improve response and recovery capabilities.

Preparing for, responding to and recovering from climate-change-fuelled disasters is becoming increasingly expensive as losses mount. All levels of government were impacted by Black Summer, and our economy took a major hit. Insurers were left reeling after the fires and then successive floods. The COVID-19 pandemic came next, and Australia was plunged into a recession.

When faced with multiple crises, it is unrealistic to expect governments to continue to foot the bill for massive recovery efforts while simultaneously trying to build our capacity and resilience for the next inevitable round of natural disasters. Insurers have always provided a means of spreading risk and aiding recovery, but increasingly they are refusing to cover assets against escalating natural disaster risks.[87]

The National Bushfire and Climate Summit convened by ELCA in June and July 2020 brought together more than 150 experts from many different walks of life and disciplines and created the Australian Bushfire and Climate Plan. The experts believed that producers of carbon pollution that is driving climate change and increasing risk and loss in the community should be made to contribute towards disaster preparation, response and recovery. It was pointed out at the summit that it is well established in law that if you produce a risk and it causes harm, you are then responsible for making good the damage and compensating those who have been affected. Such as what happened in Australia with asbestos.

When the risks associated with asbestos were well established and the impacts could no longer be denied, there was a worldwide withdrawal of the product, and manufacturers were required to pay compensation to victims.

The difference with climate change is that everyone is affected worldwide, as is every plant and animal. It is a more widespread threat than asbestos.

As we begin winding up damaging fossil fuel industries we must also find new jobs for workers as part of the transition to clean, renewable energy. Surely it is time for the fossil fuel industry to start to contribute towards government efforts to adapt to increasing climate risks and to a new economy, just as they are required to restore environments when they cease mining. A levy is not unprecedented, is well-justified, and in the view of ELCA, all government subsidies to fossil fuel industries should immediately cease, with monies used instead to fund better fire prevention, firefighting and community resilience initiatives. This may be a forlorn hope given that, at the same time in 2021, when the International Energy Agency sounded a stark warning to the world that we must stop burning fossil fuels as quickly as possible, pivot to renewables and urgently reduce emissions, the Australian and New South Wales governments announced that they would use taxpayer dollars to build new gas-fired power plants.

The immediate and ongoing task following the 2019–20 bushfire season was to assist individuals and devastated communities to get back on their feet as quickly as possible. As recognised in various inquiries, this requires a different approach to the essentially ad hoc disaster recovery arrangements that had been in place prior to Black Summer.

Recovery is a long-term, detailed and resource-intensive undertaking and we need greater long-term thinking in this space.

Pre-existing recovery arrangements in most states and territories and at the federal level generally involved the reactive appointment of an overall recovery coordinator after a major event, tasked with bringing together agencies from all levels of government as well as private sector interests such as insurance companies, to manage and expedite recovery processes.

A major problem with this approach is that learnings are often lost after each disaster. New, unfamiliar structures have to be thrown together and teams formed, lines of authority are not well understood and sometimes do not function as required, community members sometimes do not understand how to apply for help or what is available, and processes can be overly bureaucratic, involving applicants having to restate details to multiple agencies. Added to this mix is the complexity of managing and distributing charitable donations of cash and goods.

These arrangements and structures had been sufficient in the past when significant disasters were relatively rare, and of a relatively manageable size and scale. However, with the scale and frequency of natural disasters increasing, so too must the scope and capabilities of structures and processes designed to deal with their aftermath.

Both the Federal and New South Wales governments have recognised the need for a step-change, and both have created permanent agencies to manage not only recovery, but ongoing community resilience. This is a model that every state and territory should now consider.

It has to be accepted that in our lifetimes, the threat of natural disasters, including bushfires, will continue to escalate as the planet continues to warm. Capabilities to respond effectively to disasters are effective up to a point, but as demonstrated during Black Summer, the sheer scale, duration and geographic footprint can overwhelm response capabilities, meaning that help may not be available when most needed. That is why there needs to be a greater emphasis on building community resilience – the ability to take a hit and to bounce back after an impact – rather than an unrealistic reliance on outside help to make the problem go away.

There are many aspects to ongoing community resilience: actively involving and assisting communities to develop their own plans; identifying at-risk individuals and groups within the community and who will assist them; as far as possible ensuring reliability of basic services such as electricity, drinking water, sanitation and food supplies; ensuring that there are means available to alert locals to an emergency and well-practised procedures to prepare; identifying local safe refuges and triggers to use them; having local structures in place to ensure strong leadership during and after a disaster; as well as ensuring that communities are fully involved in their own recovery, an aspect essential to mental wellbeing.

There are numerous lessons learnt from Black Summer and previous disasters that can guide community resilience efforts. One of the main issues that communities faced was the loss of electrical power, creating a cascade of other failures: loss of communication and the ability to receive warnings or to call for help; loss of drinking water; loss of sewerage treatment; loss of refrigeration; loss of fuel for vehicles. All of these need to be planned for and different approaches, such as the establishment of solar micro-grids that do not rely on vulnerable telegraph poles and wires, should be part of the rebuilding.

There needs to be a nationally consistent all-hazards alert system that provides communities with targeted local warnings, whether from fire, storms, floods or other hazards. The Bushfires Royal Commission noted that while each state and territory had implemented new emergency warning processes following the 2009 Black Saturday fires, there was a lack of consistency in the types of information being provided and emergency warning icons, leading to confusion and gaps in knowledge. People near borders had to have access to two or more apps in order to obtain relevant information, noting that information was sometimes contradictory. Bushfire warning apps in some jurisdictions did not extend to other hazards, such as floods. This is an area that is receiving considerable attention from the fire and emergency services and should be relatively easy to solve.

Recent research has concluded that targeted hazard reduction aimed at creating fuel-reduced zones surrounding small towns and suburbs has the best chance of reducing fire intensity and risk during the fire seasons when fire danger levels do not exceed Severe, with the protective effect reducing significantly when fire dangers reach Extreme and Catastrophic.[88] Fuel-reduced zones will provide strategic advantage during extreme fire seasons, enabling fire services to stay and defend properties. Creating a mosaic of fuel ages across broader areas and the managed burning of ridge tops will help to slow fires and to control them when fire danger ratings are in the lower, more manageable ranges.

This will require targeted application of increased resources, particularly in land management agencies (national parks and forestry) and those urban fire services that have responsibilities for protecting life and property on the urban / bushland interface. Volunteer rural fire services will also need to play a greater role, though governments need to understand that volunteer goodwill and availability is neither a given nor a bottomless pit. For example, routinely relying on availability of large numbers of volunteers during the week could ultimately result in a loss of members, which is why the full-time agencies need to have additional resources to help fill this gap. It was ironic during Black Summer that the New South Wales Nationals Party incorrectly criticised the National Parks and Wildlife Service for allegedly doing less hazard reduction than in previous years, given that it could be argued that the Liberal–National government had presided over a situation where the Service had fewer people available to manage an expanding parks estate.[89]

Massive losses of habitat and up to 3 billion native animals highlighted that there is no coordinated approach to the rescue, rehabilitation, release and protection of native animals following major fires. A number of volunteer groups are involved in day-to-day animal rescue, but there were no formal frameworks establishing standards, training or animal welfare requirements prior to Black Summer.

In 2020, the New South Wales Government was the first in Australia to establish an accreditation framework for wildlife rescue organisations after the bushfires highlighted issues within the largely volunteer sector.[90] The framework will hopefully be used as a model nationally, as it is intended to drive the setting of standards and training requirements that will improve the ability of volunteer groups to assist distressed animals following major fires.

As the likelihood of death, injuries and property damage escalates due to worsening extreme weather, communities and governments will need to focus and improve our response capabilities. There will be increasing expectations that fire and emergency services and land management agencies are well-staffed and equipped.

While state and territory governments need to heed these expectations and meet some pressing needs, they must also be cognisant of the reality that investment in response capabilities will at some stage reach a point of diminishing returns, where additional expenditure will result in negligible improvements in outcomes.

Similar to the health system, it is well known that expenditure on proactive health improvement generates more benefits than expenditure on acute healthcare interventions. For example, programs that improve diet and exercise can reduce obesity, which in turn can significantly reduce expenditure on diabetes and heart-related diseases. Ever-increasing levels of obesity lead to increased hospital admissions and costs for acute care, taking funds away from programs that could reduce the size of the overall problem. It can be a vicious circle.

When I was fire commissioner, the three most successful initiatives in cutting the death toll from fires had nothing to do with having more firefighters and fire engines. They were early intervention measures: making smoke alarms mandatory in all residential accommodation; forcing cigarette manufacturers to produce products that self-extinguish when dropped in dry grass, on beds and furniture; and mandatory installation of fire sprinklers in aged care facilities. Escalating extreme weather will drive the need for more response resources, but this must not come at the expense of proactive strategies that could reduce long-term risk, such as continuing research, targeted hazard reduction, funding for cultural burning, improved building standards, home retrofit programs, relocating communities from unsafe locations, and community education. Many other needs will emerge.

It is a dilemma that is difficult to address, because communities tend to have a short-term view of the risk environment and want to see something immediate and tangible, such as new firefighting hardware, when fires are fresh in their memory. More intangible outcomes that can take years to come to fruition and are less understood are often overlooked or ignored. The greatest demand for the establishment of Community Fire Units came immediately following major bushfires in areas that had been impacted, which from a risk perspective did not need those resources for at least another five years. People in areas that had not experienced fires recently, where the units were most needed, were less likely to enquire.

I believe that a balanced approach to increasing response resources in tandem with investments in community resilience will create the most benefits, but only if this is on the clear understanding that changes to our climate require a massive uplift in budgets. Ultimately, no matter how much money is invested in response, it needs to be understood that emergency services will increasingly be overwhelmed by out-of-scale events driven by climate change as severe seasons become more frequent and intense. This reinforces the absolute necessity and urgency for strong government commitment to both climate change mitigation / adaptation, including deep cuts to emissions. Without this dual approach, no amount of additional emergency response resourcing will be able to keep communities safe. Emissions action is the only measure that has a chance of eventually stabilising, then reducing, the escalating risks and impacts.

When army reserves were eventually called out during Black Summer, it seemed to be a panicked, reflexive move from a defensive government that had been found wanting. It needs to be understood that the Australian Defence Force does not have the training, equipment or capabilities to directly fight fires, and any attempt to develop such a capability would simply duplicate state and territory emergency services capabilities while wasting very large sums of money. There is no plausible reason why the ADF would be deployed as a firefighting force now or in the future. It makes as much sense as asking volunteer firefighters to fight wars.

However, the ADF has a range of logistical and engineering capabilities and the ability to deploy large numbers of trained, disciplined and well-organised personnel, which can be of great assistance at disaster scenes before and after impact. During Black Summer, the ADF fulfilled crucial roles such as evacuating people by sea and air, refuelling firefighting aircraft in the field, setting up accommodation for firefighters in the field, catering, and many initial recovery tasks such as removing fire debris and fallen trees. Many of these roles released civilian firefighters, thereby reinforcing the emergency response efforts.

The Bushfires Royal Commission found, exactly as ELCA had tried to tell the Prime Minister before the fires, that processes for requesting and accessing ADF support can be convoluted and slow. It is pleasing to see that they are now being updated and streamlined so that it will be easier to mobilise resources in future.

Medium term

A range of medium-term strategies need to be implemented as quickly as possible so that over time they will serve to increase community resilience from bushfires and other natural disasters. They cannot be put off to a later date as their positive impacts are cumulative.

1. First of all, we need to start the transition away from coal. Much of the world is rapidly moving away from coal and major banks and other financial institutions including superannuation funds are increasingly abandoning fossil fuels. Australia's strong reliance on coal as an export will increasingly become a liability given that our major trading partners – including China, the USA, Europe, the UK, Japan and South Korea – have all adopted net zero emissions targets.

In May 2021, the Australian Government announced that it will commit up to $600 million of public funds to a new gas-fired power station at Kurri Kurri, in the New South Wales Hunter Valley. The announcement came on the very same day that the International Energy Agency advised there can be no new gas, coal, or oil projects if the world is to achieve net zero emissions by 2050. The federal government was yet again isolating itself from the international community. Gas is a fossil fuel. Any new fossil fuels are locking us in for further catastrophic climate impacts. Building a government-owned gas power station in the middle of a climate crisis is the equivalent of asking the Australian public to jump onto a sinking ship without a life raft.

Australia needs to urgently diversify our export offerings and recognise that the age of fossil fuel-driven wealth is coming to an end. This requires that the federal government ceases its rhetoric about protecting the relatively few coal industry jobs and create market conditions that stimulate new clean industries, new jobs, and new export opportunities. Australia has the potential to be a clean energy industrial superpower. Generations of Australians could work in clean industries such as clean manufacturing, mining, minerals processing and hydrogen made from renewables. There must be a dialogue with key unions and a program to retrain workers in a transition to a new economy and new jobs. Domestically, renewables are clearly the cheapest new form of electricity generation and market realities mean that there will almost certainly be no new coal-fired power plants in future. We should not fall for rhetoric about renewables destabilising the grid. Our national government instead needs to invest in modernising the grid and in large-scale batteries. A good place to start would be to divert funds earmarked for the planned new gas-fired power plant and from subsidies paid to fossil fuel industries.

The New South Wales Government has shown the way forward with renewable energy zones and improvements to transmission network infrastructure that enables the grid to take on more electricity from renewable sources, balanced with storage. South Australia proved that large batteries are viable and help to manage the variability inherent in solar and wind technologies. Batteries and pumped hydro systems can provide storage capacity and the ability to handle peak loads.

There need to be strong market and policy signals nationally that renewables are the way of the future, so that investors have certainty. The ultimate payoff for the community will be cheaper electricity prices, reduced emissions and a safer future.

2. Strategic hazard reduction is a key strategy to mitigate the impact of bigger, hotter, more destructive bushfires. Despite windows for conducting hazard reduction narrowing as fire seasons lengthen and the reality that fuel reduction cannot stop fires during extreme and catastrophic fire weather, creation of a mosaic of fuel ages across the landscape by coordinated strategic burning programs could eventually help lessen the impact of future extreme bushfires while protecting human lives and assets as well as natural assets. Research has established that regular burning close to homes and infrastructure is more effective than broadscale burning if protection of life and property is the objective. This will require an increase in the number of hectares of land subjected to prescribed burns each year *and* a rolling program that ensures identified areas are burnt around once every decade, with more frequent burning closer to settlements. Any such burning must be carried out in an environmentally sustainable manner that considers the fire regimes that local flora and fauna have adapted to.

Implementation of a strategic hazard reduction program presents opportunities to incorporate, in those areas where knowledge and capability already exist as well as locations where it can be readily re-introduced or learned, opportunities for cultural burning practices by Indigenous communities. There is evidence that cultural burning practices can be of assistance in managing fuel levels, and that there are likely to be other environmental advantages to the introduction of lower intensity burning. There are also sociological advantages in embracing and properly funding cultural burning, enabling Indigenous communities to develop better connection to Country and demonstrate the healing power of traditional burning methods developed by their ancestors over millennia. At present there is insufficient official recognition and understanding of cultural burning, and insufficient funding provided to make it viable.

3. The Sendai Framework for Disaster Risk Reduction 2015–2030 was adopted by the United Nations General Assembly in 2015 and provides priorities, targets and measures for countries to increase resilience to disasters. One of the four priorities is: 'Enhancing disaster preparedness for effective response and to "Build Back Better" in recovery, rehabilitation and reconstruction.'

Following the loss of thousands of buildings and the destruction of critical infrastructure for electricity, water, communications and sanitation, it is essential that long-term reconstruction considers how they were affected in the first place. For example, increased wind velocities during storms and increased frequency of serious bush-fires may require a rethink of the early twentieth-century approach of centralised power production and the use of poles and wires for all electricity transmission, given their increasing vulnerability, destruction and failure, in turn leading to escalating replacement costs. Establishment of micro-grids using renewable energy sources and batteries are likely to be a better long-term solution less likely to be affected by natural disasters, and more able to support community resilience and recovery.

As homes are rebuilt to house families rendered homeless by fires, there is an understandable desire to cut red tape and make sure the process is as fast, easy and cheap as possible. Unfortunately, such an approach can be short-sighted, particularly if homes are built in the same locations, of the same materials, and of similar design to what burned down. Local government is generally the consent authority and needs to have an eye to current and future risk. There should be no cutting corners, particularly in relation to Australian Standard 3959, Construction of Buildings in Bushfire-prone Areas. In a breathtakingly short-sighted move, in mid-2021 the New South Wales government waived certain planning requirements for people rebuilding after the bushfires, potentially rendering them uninsurable.[91]

4. Town planning also plays a crucial role in bushfire safety, with local governments in New South Wales required to take into account Planning for Bushfire Protection, a policy developed by the RFS in relation to any development on bushfire-prone land. There may now be areas formerly considered safe where there should be no rebuilding due to heightened fire risk. This is understandably an emotive, complex issue, but there is a significant precedent in situations where flooding is the predominant risk. We currently establish no-go zones for building because of the likelihood of inundation and flood damage. Insurance companies also play a role by refusing to insure in certain high-risk locations, and a reduced risk appetite by insurers has impacted many businesses and residents since Black Summer.

All of this accords with the philosophy of 'build back better'. Reconstruction should have an eye to the future and increasingly extreme fire seasons. Homes need to be capable of acting as a refuge, and of surviving a major fire.

Long term

Credible action and policies to reduce emissions

The embarrassment of being excluded from speaking at an international climate meeting in December 2020 appeared to sting Prime Minister Morrison, as did criticism from other countries of our lack of action on emissions.[92] When the newly elected US President Joe Biden hosted a Climate Summit in April 2021, the Prime Minister's speech was notable for its stark lack of ambition and a defensive tone at odds with the urgent, front-footed approach of other world leaders, and he resisted peer pressure to enter the global fold on climate action by setting clear goals. His rhetoric had progressively shifted on climate during the fires, which I hoped might signal that the Liberal / Nationals Coalition would start to heed the settled science of climate change, but the Prime Minister's international forays after the fires disappointed. Most Australians were frightened by the fires, believe that they were driven by climate change, and fear for the future. The majority want credible policies to reduce emissions, with only 12 per cent of the population supporting the government's dangerous, short-sighted 'gas-fired recovery' from the COVID-19 recession that will needlessly lock in additional emissions for decades.[93]

It appears that the Liberal–National Coalition is being held to ransom by a small, influential group strongly opposed to more ambitious climate action, apparently more influenced by the fossil fuel industry than by climate science. Former Liberal Prime Minister Malcolm Turnbull described those holding the government back from serious climate action as 'like terrorists'.[94]

It is my sincere hope that by the time this book is published our government will have adopted credible emissions reduction targets with supporting policies that, together with the majority of other developed countries that have committed to action, will help to stabilise climate risks while protecting Australia's economic future and prosperity. Climate action should be a unity ticket, not a point of difference between the major political parties. In other countries, such as Great Britain, climate action is not a point of differentiation between conservatives and progressives. Both sides of politics there recognise the scientific basis of climate change, the significant risks and impacts, and the imperative for concerted action.

In 2017, Scott Morrison, then Treasurer in the Turnbull administration, stood up in parliament, brandished a lump of coal and addressed the Speaker of the House. 'This is coal. Don't be afraid! Don't be scared! It won't hurt you, it won't hurt you. It's coal. It was dug up by men and women who work and live in the electorates of those who sit opposite.'

He taunted those – in parliament and out – who had voiced concerns that Australia's economic symbiosis with fossil-fuels and other highly polluting industries was setting us on a course to environmental catastrophe. He accused those sitting on the opposition benches of having 'an ideological, pathological, fear of coal'. He was the Prime Minister of Australia during Black Summer.

During the fires he said that, given Australia only contributes '1.3 per cent of the world's emissions' (ignoring the impact of our fossil fuel exports and that we are one of the highest per capita emitters in the world), 'to suggest that Australia doing something more or less would change the fire outcome this season – I don't think that stands up to any credible scientific evidence at all.'[95] At least one media outlet branded this as 'disingenuous', because nobody had suggested that emissions action would have an immediate effect, or that emissions could influence or cause individual fires. It underlined the difficulties ELCA encountered when trying to warn the government of an approaching bushfire catastrophe and of how climate change is driving increased bushfire risk.

It was a conversation that the government seemed very reluctant to have.

Internationally, Australia is losing respect because of our failure to join the rest of the world in committing to credible action on emissions. In October 2020, British Prime Minister Boris Johnson, a conservative, called our PM and urged him to take greater action. The public response of the PM was to repudiate any suggestion that his government should do more, while stating that Australian climate policies will be set 'here', not in or by other countries.[96] An interesting statement given that the European Union and the USA have proposed carbon border tariffs aimed at protecting their workers and industries from countries like Australia that have failed to invest in emissions reduction, thus gaining unfair economic advantage. If the government continues on its current path of climate inaction, it could have a devastating impact on our economic future.

Against the background of longstanding settled science and directly observable increases in extreme weather events – heatwaves, out-of-scale bushfires, storms and floods – many rightly feel great frustration at the lack of leadership and courage shown by some politicians globally on climate action, although stand-out conservatives like the courageous New South Wales Environment Minister, Matt Kean, give us all hope. Kean has championed renewables and battery storage in New South Wales, pointing out that it aligns with conservative values because it reduces costs for business and is good for the environment – everybody wins.

When I was in California in 2019, President Trump repeated previous claims that agencies responsible to the Democrat governor had not been effectively managing forests. This assertion failed to acknowledge that most Californian forests were under federal, not state, control, and therefore the responsibility of his own administration. The president ridiculed any suggestion that the worsening wildfire problem might be associated with climate change, despite overwhelming evidence. His solution was to concentrate on 'raking' forest floors.

During the 2020 wildfires, when a senior Californian official stated that climate change was clearly at play, citing increased temperatures, drought and extreme weather, President Trump dismissed this out of hand, assuring media, 'It'll start getting cooler. You just watch,' and asserting that scientists did not know what they were talking about. He denied requests for disaster relief funding to assist Californian fire victims until an outcry in October in the lead-up to the presidential election.

The election of President Joe Biden in the USA has changed the trajectory of US and world climate policy, as has adoption by China of a net-zero emissions target by 2060, and by other major Australian trading partners Japan and South Korea, who aim to achieve net-zero emissions by 2050. The Australian Government is becoming increasingly lonely in its lack of ambition, action and moral leadership on climate.

Coming from an emergency services background and having worked closely with the police and military for years, I am accustomed to working in critical situations under strong, ethical, decisive leaders who can assimilate and process large amounts of information quickly, make rapid, far-reaching and sometimes courageous decisions on the basis of evidence, and, if necessary, change decisions immediately when new information comes to hand. This is what the public rightfully expects from public safety professionals. It is also exactly what most people should expect from elected officials during a crisis. And we are now in a deep, potentially existential crisis.

Sadly, during the 2019–20 bushfires, leadership was found to be lacking at the national level, while state politicians such as New South Wales Premier Gladys Berejiklian displayed strength of character and decisive, evidence-based leadership. Vision, initiative and direction are required from the very top in order to set our country on a path that will lead to tangible improvements in our prospects against more extreme natural disasters.

For months, the bushfire crisis appeared to be treated by the national government more as an exercise in image-management and self-promotion than a time to step forward and protect Australians. Despite its failings during the bushfire crisis, the government seemed to rebound and was generally considered to show strong leadership in the fight against the COVID-19 pandemic, evidently learning from its mistakes, although further deficiencies in national leadership, planning and execution were exposed during the national rollout of vaccines. In the early stages of the pandemic, decisive national action was taken on the basis of the best available scientific and expert advice.

Having shown that it is capable of comprehending scientific advice, this same level of leadership and resolve must now be replicated in relation to climate change. Our government must match the resolve of every Australian state and territory government, and large sectors of the business, financial, energy and agriculture sectors in adopting a target of net-zero emissions by 2050, and preferably *well before* then. It needs to substantially strengthen Australia's emissions reduction commitments to 2030, and provide a robust process for setting regular, science-based emissions budgets thereafter. In other words, align climate policies with science and accelerating global climate action.

Every day, week, month and year of dithering and denial will simply ensure that the inevitable shifts we must undertake will be that much harder and more expensive.

The government's so-called 'gas-fired recovery' from COVID-19 will simply lock in another polluting fossil fuel for decades rather than backing what business and industry increasingly see as the answer and are seeking clear policy signals about: renewables. We need to stop this waste of our tax dollars and disregard for our children's futures.

Perhaps recognising the seismic shift in the community and the business world, the PM announced in late 2020 that the government had decided that it would no longer try to use 'excess Kyoto (or carryover) credits' to meet its obligations to cut emissions by 26 per cent by 2030, one of the least ambitious emissions reduction targets in the world. This was dressed up as a major achievement and initiative to reduce emissions, but in truth it ran against the spirit of the internationally agreed Paris Climate Agreement, and allies such as New Zealand and European countries called Australia out for considering this dubious tactic.

I liken the possibly illegal use of credits from the Kyoto agreement, signed at the international climate summit where Australia successfully argued to *increase* rather than decrease its emissions, to a football team refusing to play next week's game but claiming victory because they won the last game and want to use the points again.

As the world gradually learns to live with the virus and the pandemic era comes to an end, unless decisive worldwide action informed by science and experts is taken, climate change will continue to worsen as temperatures increase. Without strong ethical leadership, hard decisions and action, natural disasters will continue to worsen as greenhouse gas levels rise. Increasingly we will see people losing their lives, homes and livelihoods because of climate-driven extreme weather events and disasters. The impact on the natural world, such as at the Great Barrier Reef – which has lost more than half of its corals since 1995 because of climate-change-induced mass bleaching events owing to increasingly severe, frequent marine heatwaves – is incalculable.

In its final report in October 2020, the Royal Commission into National Natural Disaster Arrangements stated:

We heard from CSIRO that even under the low emissions scenario, which goes to net negative emissions, the climate does not return to a preindustrial or recent baseline type climate immediately. It takes a very long time for that to occur, and would require CO_2 to be removed from the atmosphere. According to CSIRO, it is 'more a matter of stabilising rather than returning'. Australia 'need[s] to adapt to further changes in the climate no matter what happens with emissions and we will have inevitable changes in the climate coming through for decades to come, no matter what pathway we take forward. Strong adaptation measures are necessary to respond to the impacts of climate change. Warming beyond the next 20 to 30 years is largely dependent on the trajectory of greenhouse gas emissions.[97]

Politicians are paid to lead and to make big decisions. They are not our rulers, they work for us, and we have the power to remove them if their performance is inadequate. One of the most basic duties of a national government is to ensure the safety and security of its citizens. On the existential threat of climate change, though, our government seems to have obfuscated and spun the media instead of prioritising safety and security of Australians against escalating climate and natural disaster risks.

Given the now abundantly clear evidence of worsening climate change in Australia and in other countries that are burning, it is unfathomable that a small number of politicians, despite mountains of evidence, still fail to acknowledge that we must take urgent action to lower emissions and to safeguard future generations. In April

2021, Nationals MP Matt Canavan criticised the Prime Minister for suggesting that Australia might soon adopt a net zero emissions target by 2050, saying, 'effectively this is like the 10-year-old kid trying to jump off his parents' roof thinking he's Superman; he doesn't have the technology to fly and he's going to fall flat on his face. We're going down the same path here as a country. We don't have the technology to do these things, we can't fly and we're going to fall flat on our face if we keep trying to impose these costs on jobs in regional Australia.'[98] But thankfully an increasing number of politicians, including the Prime Minister, seem to be reading the science, accepting the data, and advocating for greater action. I applaud them in standing up to people who former prime minister Turnbull described as being prepared to 'blow the joint up' if they don't get their own way, delaying climate action.[99]

While the US, UK, EU, Canada and Japan, and Australian states and territories, aim for net zero emissions by 2050, the Climate Council estimates that, on Australia's current trajectory to 2030, we will not hit net zero until 2167.

This highlights a stunning and dangerous failure in political leadership on climate, and one that is likely to irreparably damage the Australian economy as the rest of the world retreats from fossil fuels and stops buying our coal. Our government is doing very little to create new clean industries and new clean jobs, and to develop policies and programs to transition workers out of industries that will not survive. In a country that is being assailed by natural disasters of unprecedented ferocity, clearly driven by climate change, none of this makes any sense.

Continuing climate change inaction is inexcusable given what we now know and what we all saw during Black Summer. The records keep tumbling, as scientists warn that the world is now rushing towards climate tipping points that will fundamentally and irreversibly change our weather, natural ecosystems, ability to grow food, and to maintain economies.

The thirty-three members of Emergency Leaders for Climate Action are all experts in disaster planning and response and through our lengthy careers we all directly observed changes to weather, climate and risk that scientists repeatedly warned of. We now find ourselves in uncharted territory, in a new era of super-charged bushfire and natural disaster risk, fuelled by climate change, which is caused by the burning of coal, oil and gas. The only possible long-term solution is to transition away from polluting energy sources as quickly as possible, to deeply cut our greenhouse gas emissions, and to transition our economy to new clean industries that will create new jobs while safeguarding our economy and standard of living.

Afterword

WHAT WOULD THE ANZACS THINK?

HISTORICALLY, AUSTRALIA HAS OCCUPIED a position of high respect in the world, seen as a principled nation that has always done the right thing, even when difficult and unpopular, and even when sacrifices had to be made for the greater good.

I look back through recent history and am proud of some of the things our relatively small nation has achieved. In my lifetime we have taken strong, principled stands on issues such as nuclear disarmament, human rights, apartheid and terrorism. We fought in faraway wars that initially were unlikely to directly affect Australia.

I think of my grandfather, Philip Mullins, who fought on the shores of Gallipoli and was later shot in the face fighting on the Western Front in Belgium during World War I. Of my dad, Jack Mullins, who signed up to the Royal Australian Air Force in World War II, prepared to die for his country if necessary. Of my father-in-law, Doug, who saw horrific hand-to-hand combat in the Pacific and was damaged by his experiences for the rest of his life.

Together with hundreds of thousands of other Australians, they each made personal sacrifices in support of an ethos they firmly believed in. Our nation participated in far-flung war efforts because Australians believed in making sacrifices to secure a safe future.

Imagine if, back then, a politician had stood up in parliament and said, 'We only contribute a tiny portion of the effort, so what we do makes no difference'; or, 'I don't believe that we're actually at war and I don't believe what the generals are saying.' Would the prime minister of the day have given them tacit support by saying, 'They are entitled to their opinion,' just as ours has on a number of occasions when ill-informed politicians espoused their extreme views on climate change?

I know that my grandfather, father, father-in-law and other ANZACS would have been appalled, just as all of us should be now.

Individuals can make a difference. It is up to each and every one of us to make sure that this terrible story eventually has a happy ending. We can't afford to lose hope, even though the continuing political inaction on climate change makes maintenance of a positive outlook difficult. The happy ending that we need is a global political turnaround that results in urgent action to reduce the emissions that are driving worsening extreme weather events, natural disasters, death and destruction. I'm encouraged by the political changes in the USA that have put climate action back on the agenda, with science and fact replacing conspiracy theories and uninformed opinion.

Collectively, we can't afford to let the forces of ignorance and greed continued to delay or stop what might become a fight for survival if worldwide emissions continue to increase rather than rapidly reach zero. If the almost 3 billion animals and nearly 500 people that perished as a result of the Black Summer fires could talk, they would plead with us to take immediate action. If future generations who will face even worse fires and natural disasters could talk to us, they would *demand* that we act on their behalf.

I hope that people who read this book will learn things that will help to keep them and their families safe. I hope that they will gain an understanding and realisation that climate change is no longer a theoretical concept or something that might happen in future, or that there isn't any lack of clarity about why it is happening: it is here right now, and it is threatening our safety and security. Worsening bushfires are a symptom of something far worse that will have far-reaching, increasingly deadly consequences.

Hopefully I have been able to explain through these pages the simple, settled science of climate change and, in doing so, have planted questions in people's minds about why some politicians and sections of the media are either not doing their homework, lack the intelligence to understand simple concepts, or are stretching the truth. This might raise questions in some people's minds about their fitness to continue leading and representing us.

We face a stark and increasingly obvious choice. We can easily just sit back and allow the political climate wars in Australia to rage on, a blood-sport with the inevitable result that future generations will face natural disasters, death and destruction on a scale that we can't imagine, despite having endured our shocking Black Summer. The Great Barrier Reef is in serious peril: it can't survive the relent-less warming of the oceans, and at 2 degrees of warming, 99 per cent of corals will be dead – gone forever. Warming has already reached 1.4 degrees in Australia. I don't see sufficient outrage about this, considering not only that the reef is a World Heritage icon support-ing abundant marine life, but also that it contributes $6.4 billion every year to the Australian economy and provides employment for more than 64,000 people. It is time for us to say enough is enough, and to demand that Australia transforms from being a laggard to a leader on climate and emissions action, given that we are on the leading edge of the worst impacts.

As one of the most exposed countries in the world to the increasingly dire consequences of global warming, all Australians have an urgent moral duty to our children and to our grandchildren to do everything we possibly can to keep them as safe as possible. There is no question whatsoever that policies and actions adopted today will directly impact on their tomorrow. If these policies and actions successfully drive down greenhouse gas emissions, fire and disaster risks may stabilise and eventually decline. If we continue on our current path instead – an emissions policy vacuum and a dangerous, immoral game of Russian roulette with our kids' futures – then conditions will inevitably become even more dangerous.

Many of us will still be around in 20 or 30 years, a time that the Bushfire Royal Commission warned will either see fire and disaster risks stabilising as a result of resolute worldwide climate action, or, alternatively, transforming into a nightmare scenario. If my message makes you upset, concerned, maybe even angry, and then you start to think that it might be a good idea to demand more action on dangerous emissions, then I will have done my job.

If I am still around then and my grandchildren and great-grandchildren ask me the question, 'What did *you* do about action on climate change after the catastrophic bushfires in Black Summer 2019–20?' I will have an answer. So too will my colleagues in ELCA, the Climate Council, climate scientists, stand-out poli-ticians like Zali Steggall and Matt Kean, and other organisations and individuals currently being ignored or ridiculed by the federal government. A number of current government politicians will also have answers, but they will be quite different to mine: indeed, they may not feel comfortable answering, because the impacts they denied and sometimes ridiculed will be there for all to see. They will be found to have been sitting firmly on the wrong side of history, with their incredible negligence, arrogance and lack of care part of the historical record.

I hope that everyone takes time to ponder that crucial question: *What will I tell my grandchildren, and will I be able to say that I did everything possible to keep them safe?*

Appendix

THE FIRE DANGER INDEX AND FACTORS AFFECTING FIRES

THE STANDARD TOOLS USED in Australia to predict fire danger are the Forest Fire Danger Index (FFDI) and the Grassland Fire Danger Index (GFDI), both developed by Alan McArthur. This will change as researchers are currently trialling new fire danger tools.

Fire danger calculations presently use the variables of number of days since rain, rainfall to 9 am, drought factor (based on the previous variables, a number from 1 [wet] to 10 [very dry]), lowest forecast relative humidity, maximum temperature and highest wind speed.

Fire danger indices are calculated on a scale of 1 to 100. Theoretically, at a fire danger index of 1, fires will not start or burn. At an index of 50 and above, fires will burn so intensely that they can be incapable of being controlled. After the 2009 Black Saturday bushfires, the new fire danger rating of Catastrophic ('Code Red' in Victoria) was added to recognise the increasing incidence of off-the-scale (over 100) fire danger indices, where lives are likely to be lost.

Due to the influence of climate change, fire weather conditions have progressively worsened over the years. In February 1983, during the Ash Wednesday fires in Victoria and South Australia, off-the-scale fire danger indices over 100 were recorded. Once considered to be rare, the scale was again exceeded during the 2003 bushfires in Canberra, the 2009 Black Saturday fires in Victoria where indices of more than 200 were recorded, the 2013 bushfires in Tasmania, and 2017 bushfires in New South Wales. It happened regularly over Black Summer 2019–20.

The original McArthur Fire Danger Index was as follows:

Forest fire danger index	Fire danger rating
0–5	Low
5–12	Moderate
12–24	High
24–50	Very High
50–100	Extreme

After the Black Saturday fires in 2009 it was revised to:

Forest fire danger index	Fire danger rating
0–12	Low to Moderate
12–24	High
24–50	Very High
50–75	Severe
75–100	Extreme
Over 100	Catastrophic (Code Red in Victoria)

THE FIRE PENTAGON

BASIC TEXTS DESCRIBE THE fire triangle: fuel, oxygen, and a heat source, a good way to introduce the basic chemistry and physics of combustion. Bushfires rely on these and other factors.

If it is assumed that there is a plentiful supply of oxygen (always the case outdoors), of fuel (dry vegetation) and heat sources (lightning, escaped campfires, fallen power lines, arson etc.), there are at least two other major factors that need to be considered in relation to bushfires: weather and terrain.

Fuel

There are different types of fire fuels in the landscape. Fine 'flash fuels' in grasslands dry out readily, are quickly and easily heated by the sun, and are rapidly preheated by radiation and convection from flames during a fire. That is why grass fires can start easily, build up rapidly and spread faster than a bushfire, but the heat output is lower because there is less fuel overall.

Bushfire fuels vary depending on the type of vegetation, but generally include fallen sticks, twigs, leaves, bark, native grasses and small shrubs. 'Available fuel' refers to vegetative matter 6 mm or less in diameter (about the thickness of a pencil), because this is what fuels the intensity of fast-moving flame fronts. Anything thicker burns once the fire front moves past.

Fuel loads are measured in tonnes per hectare and vary significantly based on types of plant communities, the weather, time since the last fire and other variables such as drought.

During drought, some grasslands die off and there can be insufficient fuel to carry a fire. In a forest, however, drought can increase the amount of available fuel, because the fuel bed is drier and some trees start to shed bark, leaves and even small branches as they become stressed from lack of water. This can add up to 7 tonnes per hectare to the fuel load over a short period.[100]

How fuels are deposited and arranged makes a difference to how easily and rapidly they will ignite and burn. If they are tightly packed, for example packed-down casuarina needles, they will not burn as readily because there is less oxygen surrounding the fuel. If lower layers are damp, the higher moisture content will further restrict the amount of fuel that can burn.

If fuels are more aerated or loosely packed, they will burn more readily. For example, a forest floor where leaves, bark and twigs have fallen from large trees and are suspended by grasses, shrubs and fallen tree branches promotes drying and rapid burning. When a fire starts, there is a plentiful supply of oxygen surrounding the suspended, aerated fuels, so they burn more readily than if they were compacted.

Living trees and woody shrubs over a certain diameter generally don't burn due to their sap and moisture content, but their leaves can contribute to fire intensity during hot, dry conditions. Many native trees have leaves that contain volatile oils and resins, such as eucalyptus and tea tree oil, which burn readily.

Grassland fuels are quite different to forests. There are two basic types of natural grassland fuels: annual and perennial grasses; and commercial crops such as wheat.

Annual grasses have a short life cycle and dry out naturally as the warm season progresses. Perennial grasses have deeper roots but also dry out annually. The drying process is called curing and is essentially irreversible. Even if it rains, annual grasses do not regenerate once the process of curing has progressed past a certain point (about 60 per cent cured) although seeds in the soil may germinate, leading to some new growth that remains green for a short time. The drying of perennial grasses can be delayed if there are good falls of rain, but if there is no rain they will also fully cure. The colour of grassland is significant: yellowing reflects reducing moisture content and how readily the grass can be ignited and burn.

Fuel loads in grasslands are much lower than forested areas, meaning fire intensity is lower, but fires start more easily and spread faster because light fuels are influenced far more quickly and readily by weather conditions and sunlight.

Oxygen

The process of combustion mostly involves oxygen bonding with carbon and hydrogen atoms in a reaction that produces heat and light (flames). Bush and grassland fuels mainly comprise cellulose ($C_6H_{10}O_5$) and when burned produce carbon dioxide, water vapour and heat. Fires inside buildings can be different because of the variety of fuels (plastics, synthetics, etc.), and also because the oxygen supply can be limited inside compartments. When there is insufficient oxygen present, carbon monoxide (CO), a poisonous flammable gas, can be produced as well as carbon dioxide. Carbon monoxide is lighter than air and rises, meaning that it does not build up at ground level outdoors where it is subject to wind and convection currents. It rarely forms in significant quantities if there is plentiful oxygen available, so is normally not of great concern to bushfire fighters. The exception can be if there are underground peat fires or fires involving coal stockpiles. In these situations, firefighters need to have respiratory protection (self-contained breathing apparatus, as filter masks are useless) and limit exposure, as carbon monoxide bonds easily with red blood cells and excludes oxygen, a situation that is not easily reversed and with sometimes fatal consequences.

For the purposes of understanding bushfires it can be assumed that there is always a plentiful supply of oxygen outdoors because it makes up just under 21 per cent of the atmosphere. It is quickly replenished by the inrush of air when combustion takes place and by wind currents.

Heat

For a fire to start there needs to be a source of heat capable of raising the temperature of a combustible material above its ignition point (e.g. paper, about 350°C) so that self-sustained combustion can take place. Solids have to be heated to a temperature at which they produce flammable vapours in a process called pyrolysis – the vapours burn, not the solid itself. If fuel is wet or has a high moisture content, more energy is needed to dry it out before energy becomes available to raise the solid to its ignition temperature.

In the open landscape the ability of fuels to ignite is very much related to weather, and particularly moisture and rainfall. The most common natural source of ignition for a bushfire is lightning. Obviously, though, if lightning strikes a tree during a storm accompanied by heavy rain, even if it initially catches fire, it will usually be quickly extinguished. On the other hand, if a dry lightning storm that produces little or no rain happens during a drought, the possibility of fires occurring when there are lightning strikes is greatly increased.

Heating by the sun plays a large role in bushfire behaviour and influences how readily fires start and spread. Fuels go through a daily cycle of heating and cooling owing to the movement of the sun. More importantly, solar radiation has a significant effect on fuel moisture content, with fuels drying out as they are heated. Gardeners know that during the day their plants can wilt because of the drying effect of the sun. Bush and grassland are no different and by late afternoon on a hot day fuels can be very dry. This is one of the reasons why fire behaviour is often more intense in the afternoon.

At night in the cooler, moister conditions after the sun has set, fuels absorb moisture from the air as humidity increases, making it harder for them to ignite.

During a bushfire, heat is given off by flames and is transmitted by radiation, convection and, to a limited extent, conduction. Radiation and convection preheat fuels, bringing them closer to their ignition temperatures while also drying them out. These effects combine to make fuels ignite and burn more readily.

Weather

If fuel, oxygen and heat conditions enable a fire to start in bush or grassland, weather becomes the major determinant of how it will behave and whether it will be easily extinguished, self-extinguish or rage out of control.

Rainfall

If forests and grasslands are wet, they will not ignite or sustain burning. If a fire is able to start in damp conditions, it may self-extinguish without any need for intervention. Even on a dry windy day following rain, a fire might start while the sun is out but after sunset it is likely to self-extinguish as moisture levels in the air increase.

Fuels absorb moisture from the soil as well as from the air. At night, humidity levels usually increase, and fuels absorb the additional moisture from the air, requiring more heat energy to drive out moisture before they can ignite. Wetter lower layers of fuel will transfer moisture to the layers above once the heating and drying influences of the sun and wind dissipate. For these reasons and because it usually becomes cooler and less windy, fire intensity typically declines as night progresses.

Firefighters keep a close eye on dryness throughout the year using the Keetch-Byram Drought Index (KBDI) and sometimes also Mount's Soil Dryness Index (SDI) which give indications of current rain effects on fuel and soil dryness. Percentage of curing (drying out) of grassland is another indicator of dryness.

The KBDI is a simple water balance calculation that enables moisture content to be estimated in the top 20 cm or so of soil, and therefore in fuel beds. Daily observations of temperature, relative humidity, rainfall and number of days since rain are recorded at hundreds of sites and a formula is used to calculate an index between 0 and 200. A KBDI above 100 indicates extreme dryness. Fires will spread rapidly and are likely to generate spot fires because fuels are critically dry. A figure below 25 means that there is significant moisture in the soil and fuel bed, and fires will be difficult to ignite and sustain. The KBDI is used to calculate a drought index from 1 to 10 when calculating the Forest Fire Danger Index and Grassland Fire Danger Index. A drought factor of 1 indicates that fuels are moist and only 10 per cent of the fuel is available to burn, while a drought index of 10 means 100 per cent of fuel is available.

Drought and flash drought
Long-term rainfall deficiencies can result in drought. Severe drought can influence the amount of available fuel because even heavier fuels will become critically dry, requiring less energy to catch fire and stay alight. Even a small increase in available fuel, for example an increase from 6 mm diameter to 8 mm, can significantly increase fuel loads and therefore fire intensity. Increased fire intensity can

result in more convection, more spot fires, faster rates of spread and greater difficulty gaining control.

Studies have shown a link between dryness and the number of fires started by lightning. There are suggestions that a warming climate may increase the number of dry lightning storms and further research is being conducted to explore this. There is a clear linkage between increased lightning ignitions and fuel dryness – the drier the fuel, the more likely it will ignite and cause a bushfire following a lightning strike.

Flash droughts or heatwave droughts were first recognised in about 2001 and recognise the effects of very hot weather and associated evaporation leading to drought-like conditions, but without the long build-up. Drier soils lead to low fuel moisture content and higher fire risk.

Evaporation

Evaporation is greatest when temperatures are high, humidity is low, winds are strong and there is abundant sunlight. With climate change impacting year-round temperatures both at night and during the day, rates of evaporation can increase significantly. When coupled with long-term trends of reduced rainfall in some locations this can have a significant effect on fuel and soil moisture, and the severity of droughts when they occur.

Relative humidity

Relative humidity is a measure of moisture in the air. High humidity means there is more moisture in the air, whereas low humidity means that it is drier.

Humidity influences fires through the uptake of moisture by dead and living vegetation. Throughout the day, dead leaves, twigs and bark absorb or lose moisture from the surrounding air, affecting their fuel moisture content and flammability.

When humidity levels are very high, it can be difficult to ignite fuels even if it hasn't been raining. Fires won't burn as intensely because heat energy is needed to drive moisture out of adjoining fuels so that the fire can continue to spread. Spot fires are less likely because moisture in the air makes combustion difficult to sustain as pieces of bark and leaves drawn up into a smoke column travel through the damp air.

When relative humidity drops, fire behaviour can suddenly worsen. Relative humidity below 25 per cent can lead to dangerous conditions, and firefighters look out for pulsing flames and black smoke as an indicator that the speed and intensity of combustion is increasing due to reduced humidity.

Fuel moisture content is directly related to humidity: lower humidity levels draw moisture out of dead and living fuels, meaning that less energy is needed to bring them to their ignition point and to sustain burning. Spot fires become far more likely because dry air makes it easier for burning embers in the air to stay alight until they land in unburnt bush, sparking new fires.

Wind

Of all the weather parameters that affect bushfires once they ignite, wind is the most significant. Speed and direction of the wind are critical elements in determining how a fire is likely to spread.

Different parts of the country are affected differently by wind direction. The main determinants of whether a particular wind will intensify fire behaviour is whether or not it is coming from the dry inland, and also whether or not it is coming from the warmer north. As weather systems transit the country, large bodies of air are heated and dried as they travel across the often hot, arid centre of Australia. Research has also identified hazards associated with dry slots of air in upper atmospheric layers that under certain conditions can mix down to the surface, changing fire dynamics.

In Western Australia, an easterly wind can be very dry, and in New South Wales a westerly wind is the driest. In both cases this is because the winds are coming from the dry centre of Australia. If hot air is drawn down from the north, winds will be warmer. If air is drawn up from the south, it will be cooler. Therefore, the hottest, driest winds that often drive major bushfires are easterly and north-easterly winds on the west coast, and westerly and north-westerly winds on the east coast. In Victoria and South Australia, northerly winds are often problematic, and in the Northern Territory, winds from the south can cause elevated fire danger.

Winds that have a southerly influence will be cooler, and they will pick up moisture if they move across the ocean. Therefore, in Western Australia southerly and westerly winds, and in New South Wales southerly and easterly winds tend to bring cooler, moister conditions and possibly rain.

Wind speed significantly affects fire behaviour. Strong winds do several things to a fire: they bend flames forward so that fuels are preheated and therefore burn more readily; they increase oxygen supply to the fire, increasing the rate of combustion; and they carry embers ahead of the fire, starting new spot fires.

Sudden wind changes after a day of serious fire weather are a significant hazard. For example, a strong southerly wind change can instantly transform the lower intensity northern flanks of a fire into a high-intensity fire front. Many firefighters have been killed or injured and many buildings destroyed when a major fire has been affected by a significant wind change, because it immediately drives the fire in a different direction. This is particularly the case in Victoria where bad fire days usually involve hot, dry northerly winds. Cool fronts with strong south-westerly winds often arrive late in the day, turning the northern and eastern fire flanks into high intensity fire fronts.

On the east coast, where the main fire winds are from the west and north-west, strong wind changes usually come from the south, together with a cool front. That is why New South Wales firefighters always concentrate their efforts on trying to contain and secure the northern flanks of fires when a wind change is forecast, and they may be withdrawn or ordered to take shelter when the front arrives.

Even when a wind shift results in the temperature dropping and humidity increasing, there will be a lag effect that results in fires continuing to burn intensely for an hour or more. This is because dead and living vegetation take time to absorb the increased moisture in the air and to lose heat to the atmosphere as air temperature drops.

Atmospheric stability

Atmospheric stability refers to the ability of parcels of air to move upward or downward through atmospheric layers. If conditions favour the mixing of different layers of air, the atmosphere is said to be unstable. On a normal day this can result in the formation of storms. If air masses are more likely to remain at the same altitude and not mix, the atmosphere is said to be stable.

Atmospheric stability can have a significant impact on fire behaviour and on the likelihood of pyrocumulous clouds (fire-generated storm clouds) forming from a smoke column. In extreme conditions the pyrocumulous clouds can develop into pyrocumulonimbus (pyroCb), or fire-generated thunderstorms that can generate extreme winds, extreme fire behaviour and dry lightning that can start new fires long distances from the original fire.

PyroCb are the most intense and dangerous example of fires coupling with the upper atmosphere. They have been associated with large-scale damage and deaths, such as in Canberra in January 2003 when 487 homes were destroyed and four people were killed, and Kinglake in Victoria in February 2009 when 159 people were killed and hundreds of buildings were destroyed.[101] PyroCb can lead to winds suddenly changing direction, violent downdraughts, cyclonic wind velocities, fire tornados and dry lightning that can start new fires kilometres ahead of the main fire.

The continuous Haines (cHaines) Index is used to predict unstable atmospheric conditions that can contribute to the development of pyroCb events. The cHaines Index combines measures of the vertical rate of change in air temperature and the change in moisture content of the lower atmosphere to provide a score out of 13. A high cHaines Index value indicates an unstable atmosphere that can favour pyroCb formation.

Unstable air masses allow convection columns to form readily over fires and to reach high altitudes. This means that fires then draw in large amounts of air at the base, just like air being drawn into a chimney.

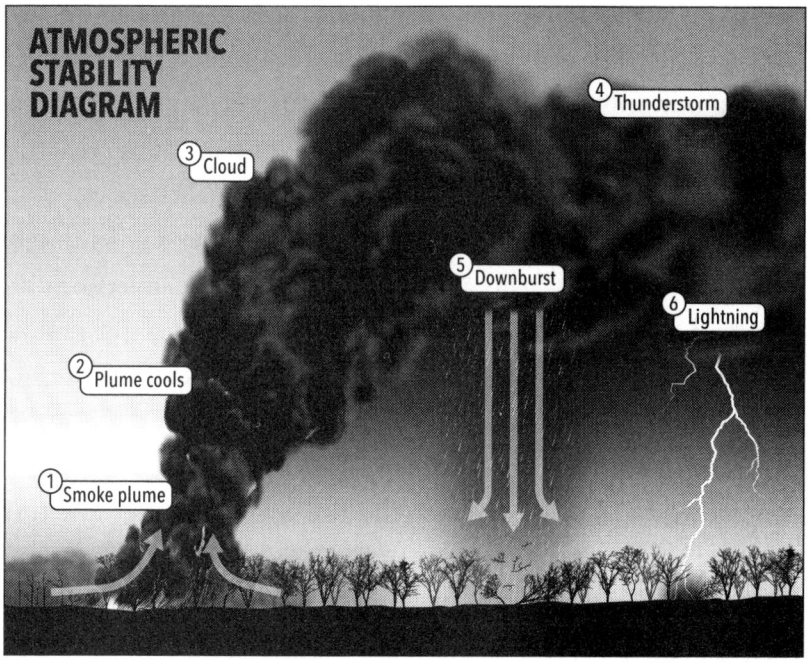

ATMOSPHERIC STABILITY DIAGRAM

3 Cloud
4 Thunderstorm
5 Downburst
6 Lightning
2 Plume cools
1 Smoke plume

The inrush of air at the base can increase wind velocity at ground level, lead to more intense burning and draw burning embers high up into the convection column, resulting in long-distance spotting. There can be strong downdraughts and indraughts of air at ground level to replace the air being drawn into the convection column. Air that is descending from high altitudes to replace air sucked into the fire at ground level can be compressed and heated as it gets closer to the ground. If air at upper levels is very dry, downdraughts can significantly change fire behaviour as more moist air near the ground is replaced by descending warmer, drier air. Upper-level winds are usually stronger than at ground level, and a coupling between the fire and the upper atmosphere can bring these stronger winds down to the ground, sometimes forming fire tornados and fire whirls.

In stable conditions there is little vertical mixing of air, and the development of convection columns is impeded. Temperature inversions can form, meaning that a stable layer of warm air overlays a cooler layer. The warm layer acts like a blanket and stops air from mixing upwards. An inversion above a fire can impede development of convection columns, reducing fire intensity. It can also trap smoke at ground level causing visibility and air quality issues.

Atmospheric stability changes throughout the day and night as the atmosphere warms and cools. Temperature inversions are common at night and in the morning with cool air near the ground unable to break through layers of warmer air above. As the sun rises it warms the land and lower air layers and the temperature inversion eventually breaks down. When it does, fire behaviour can suddenly increase because the fire may form a convection column, but also because stronger upper-level winds start to move down towards the ground. That is why firefighters monitor forecasts closely to determine when the surface inversion will break, and when upper atmospheric layers will mix down to ground level, bringing stronger winds.

Terrain

Fires are affected significantly by landforms and geography. Factors that are considered by firefighters include degree of slope, the aspect of the land (direction it faces), and altitude.

Fires move faster uphill than downhill. If a fire starts in a flat area of uniform fuel type and there is no wind, the fire will generally burn in a roughly circular pattern at a similar intensity all around the perimeter. Introducing even a slight slope changes the fire dynamics considerably.

The rate of fire spread doubles on a 10-degree slope and quadruples on a 20-degree slope. If a fire is burning downhill it can decelerate by a similar factor, which is why firefighters always avoid being upslope from a fire and often base fire attack plans on tackling fires when they start to burn downhill.

The simple physical relationship here is that flames and heat both rise. On level ground the main preheating of fuels is caused by radiation from flames. As a fire starts to burn uphill, flames bend closer to fuels above, increasing exposure to radiant heat; and convection also starts to play a part in preheating fuels. That is why fires usually accelerate and intensify very quickly when they start to burn uphill.

The direction land faces plays a large role in the flammability of fuels and how fires behave. Slopes that are exposed to sunlight for most of the day are drier and the temperature of fine fuels on them will be higher because of heating by the sun. Aspects most exposed to the sun are also usually more exposed to hot, dry winds.

On the east coast, slopes that face north, north-west, west and south-west are generally more fire-prone and problematic for firefighters because they are exposed to the sun for most of the day, and to warm, dry westerly, north-westerly and northerly winds that reinforce drying. On the west coast, the opposite applies, with the driest winds coming from the east.

Because (on the east coast) wet weather and moist air often come from the south and east, the vegetation on south, south-east, east and north-east facing slopes often have a higher moisture content and can therefore be less flammable. They are less likely to be directly affected by hot, dry winds, so fires may not generally burn as intensely or quickly as on the more exposed aspects (during drought and very hot weather, however, this may not be the case).

It was observed following the 1994 New South Wales bushfires that many south- and east-facing slopes burned severely. Subsequent research identified that during extreme conditions and when fuels are very dry, there can be little difference in fire intensity related to aspect. Mass spotting and vorticity-driven fire spread can result in the usually less-hazardous aspects also experiencing extreme fire behaviour. This is worrying given the increased temperatures and intensification of fires due to climate change.

Winds can be stronger in upper levels of the atmosphere than near the surface. This means that elevated mountainous areas can experience stronger winds, often earlier in the day, than locations at lower elevations. Higher elevations are also less likely to be protected by surface inversions for as long as locations closer to sea level. For example, during serious fire weather, Katoomba in the Blue Mountains often experiences wind velocities higher than in Sydney at sea level. Stronger winds combined with steep mountainous slopes and heavy fuels can make firefighting extremely difficult and hazardous. Research has shown that pyroconvective fire conditions are most likely to occur in rugged, elevated locations likely to experience intense fires.

FIRE DYNAMICS

THE PREVIOUS SECTION PROVIDES a basic understanding of how fuel, heat, oxygen, weather and terrain affect fires. This section explains in more detail how fires behave once they start.

Ignition and transition

Regardless of the mechanism of ignition, fires develop and build up in a similar manner and are affected by fuel type, amount and distribution, dryness, weather and terrain.

The initial stage of a fire requires a heat source that will raise the combustible material to its ignition temperature. The fire then goes through a build-up period where it preheats fuels around it and starts to spread slowly. In benign weather conditions, the fire will take time to build momentum. It will spread out in all directions fairly uniformly unless influenced immediately by slope, wind or availability of dry fuels. Fuel moisture plays a significant role in the early stages. In moist fuels a fire might self-extinguish because it cannot produce enough heat energy to drive moisture out of adjoining fuels.

If a fire does not self-extinguish, it will progress until it reaches a stage where it starts to build momentum: the transition stage.

On reaching transition, flames tend to lean forward towards unburnt fuels, and heat from the fire causes air flows to change, increasing wind strength in the immediate vicinity and accelerating the fire. Depending on weather, fuels, and topographical conditions, the fire may then continue to develop until it forms a convection column and starts to generate spot fires, greatly increasing its rate of spread.

The exception is grass fires which are more likely to be wind-driven than driven by a convection column. High winds can prevent a convection column from forming above a grass fire and once wind reaches a certain velocity the rate of spread of a grass fire will stop increasing, even if wind speed continues to increase.

The amount of time taken to go through the various stages varies with location, fuel types and weather. In serious fire weather, fires can build up extremely quickly, which is why rapid response is critical on the worst fire weather days.

Spotting

During a bushfire, burning bark from trees, small sticks and leaves can be carried up into the convection column, then be carried by the wind until they fall to the ground and start new fires at a distance from the original one. These new fires are called spot fires.

Some species of eucalypt trees have bark that can stay alight if drawn up into a smoke plume. The worst types are called candle bark: long streamers of smooth bark that curl up and hang from tree branches. When a fire burns below, the candles catch fire and are easily drawn upwards into the smoke column by convection. The shape of candle bark makes it aerodynamic, so it can stay alight for some time while being carried high into the air. Spotting distances of up to 30 kilometres have been recorded when there are strong upper-level winds, and during Black Summer spotting distances of 8–12 kilometres were common on the worst days.

Fibrous stringy-bark trees are very flammable and can generate intense short distance spotting up to 5 kilometres ahead of the main fire, with intense spotting to about 200 metres common.

Where mass spotting occurs, the many new fires can coalesce and result in large areas of flaming that are difficult and dangerous to attack. They can change local wind conditions due to inflows of air, then themselves generate even more spotting.

Spot fires are more likely when fuels are very dry, wind velocity is high, there is an uphill run of fire, fire reaches into tree crowns, or where there are very heavy fuel loads. They are dangerous and unpredictable and make the work of firefighters especially difficult because it is much harder to predict fire spread if there are multiple fire fronts. Fire crews can be easily outflanked by unexpected, rapidly developing spot fires.

Junction zone

When two fires approach and then join, fire intensity increases significantly at the junction point, often generating significant spotting. As two or more fires get close to each other they start to become affected by fire-influenced air currents as air is drawn into the base of each fire. Flames will bend towards each other and fires will start to draw together.

As two flame fronts approach each other, the drying and heating effect on fuels is amplified because radiation and convection are now coming from two or more directions. This rapidly raises the temperature of fuels closer to their ignition temperatures, moisture is driven out, and as a result flame heights and fire intensity can increase rapidly. As two fires meet, large flames and significant heat can be generated. This is always a consideration during backburning operations – if there is insufficient depth to a backburn when it meets the main fire and produces a surge in intensity, spot fires can cross control lines.

El Niño and the Indian Ocean Dipole

The El Niño Southern Oscillation (ENSO) is a global weather system that has exerted significant influence on bushfire weather in Australia. It refers to sustained warming (El Niño) or cooling (La Niña) of central and eastern Pacific waters. Ocean temperatures have a large influence on weather systems affecting Australia and in particular on rainfall.

El Niño years are typically hotter, drier and more likely to produce serious bushfires, while La Niña years are typically cooler, wetter and more likely to produce tropical cyclones. During an El Niño the cooler ocean waters close to Australia produce fewer clouds and less rainfall, and east coast temperatures are likely to be higher.

An El Niño had previously been considered a prerequisite for a serious bushfire season in New South Wales, yet it was absent in 2013 and during Black Summer. It has been established that global warming is starting to swamp the influence of some climate drivers.[102] The World Meteorological Organisation has reported that La Niña events, noted for bringing cooler weather, increased rainfall and cyclones to the east coast of Australia, are now hotter than El Niño events were in the 1980s.[103]

Another major driver of Australian weather linked to serious bushfire seasons in Australia is the Indian Ocean Dipole (IOD). The IOD is a measure of ocean temperature to the north-west of Australia. A positive IOD causes waters in the eastern Indian Ocean to be cooler, reducing evaporation and moisture in the air. A positive Indian Ocean Dipole can therefore worsen fire weather as there can be less rainfall in winter and spring. The strong positive IOD in 2019 did not break down until January 2020. The BoM explained how the strong positive IOD helped set the scene for Black Summer:

Australia's climate was impacted by a positive Indian Ocean Dipole (IOD) in both 2018 and 2019, exerting a drying influence over many parts of the country. The IOD is characterised by cooler waters to the north-west of Australia and warmer waters further west towards Africa. The positive IOD contributed to low rainfall over southern and central Australia during winter. It is unusual but not unprecedented to have successive positive IOD events (based on the Bureau of Meteorology's criteria, the last known occurrence of consecutive positive IOD events was in 1982 and 1983). While the IOD is a natural mode of variability, its behaviour is changing in response to climate change. Research suggests that the frequency of positive IOD events, and particularly the occurrence of consecutive events, will increase as global temperatures rise.[104]

Throughout history, some of the most serious bushfires have occurred when a strong El Niño coincided with a strong positive IOD, resulting in sustained hot, dry conditions, exacerbating periods of serious fire weather. Examples of this include Ash Wednesday in Victoria and South Australia in 1983, and Black Saturday in Victoria in 2009.

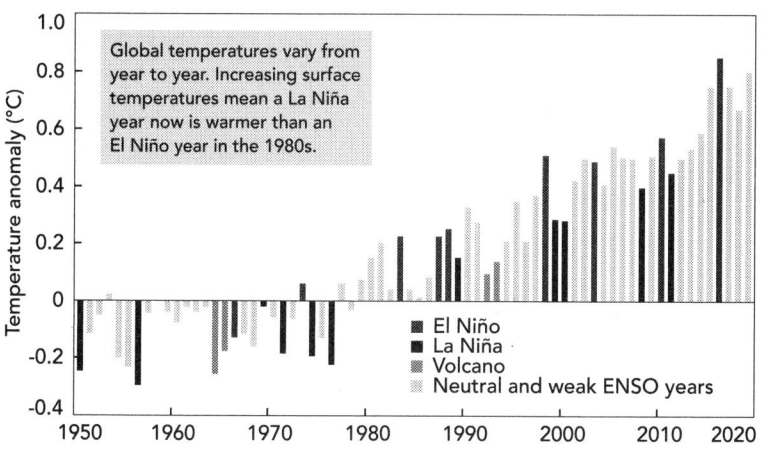

Sudden stratospheric warming and the southern annular mode

The Southern Annular Mode (SAM), influences the north–south movement of a band of strong westerly winds that circulate near Antarctica below Australia. A negative SAM in summer results in the band of winds moving further north and can result in hot, dry westerly winds affecting the east coast. A negative SAM is also associated with reduced rainfall in south-eastern Australia and increased fire dangers in New South Wales and southern Queensland.[105]

Sudden warming of the stratosphere over the South Pole from August 2019 induced a negative phase of the Southern Annular Mode.[106] It was the strongest warming event since 2002.[107]

Coupled with the strong positive IOD, the SAM contributed to the record hot, dry conditions in 2019.[108]

A recent study found that the SAM is heavily influenced by 'anthropogenic changes in greenhouse gas concentrations and stratospheric ozone', and as a result has become more positive over recent decades, contributing to lower rainfall over the south of Australia and, as a result, increasing levels of bushfire risk.[109]

ENDNOTES

1 Sanderson, B.M. and Fisher, R.A., 'A fiery wake-up call for climate science', *Nature Climate Change*, vol. 10, 2020, pp. 175–77; doi.org/10.1038/s41558-020-0707-2

2 Doogan, M., 'The Canberra Firestorm: Inquest and Inquiry into four deaths and four fires between 8 January and 18 January 2003, Volume 1', ACT Coroners Court, 2006; courts.act.gov.au/__data/assets/pdf_file/0007/961468/canberra_firestorm_vol_i.pdf

3 Cruz, M.G., Sullivan, A.L., Gould, J.S., Sims, N.C., Bannister, A.J., Hollis, J.J. and Hurley, R.J., 'Anatomy of a catastrophic wildfire: The Black Saturday Kilmore East fire in Victoria, Australia', *Forest Ecology and Management*, vol. 284, 2012, pp. 269–85; sciencedirect.com/science/article/abs/pii/S0378112712001223

4 Ibid.

5 Owens, D. and O'Kane, M., 'Final Report of the NSW Bushfire Inquiry', Parliament of NSW, Sydney, 2020

6 Wilson, A. and Ferguson, I., 'Fight or flee? A case study of the Mount Macedon bushfire', *Australian Forestry*, vol. 47, 1984, pp. 230–36, 10.1080/00049158.1984.10676007

7 Dillon, H.C.B., 'Inquiry into fire at Wambelong Camp Ground, Warrumbungles National Park, New South Wales January 2013', State Coroners Court, Glebe, 2015; coroners.nsw.gov.au/documents/findings/2015/Warrumbungles%20findings%20Final%2028%2008%2015.pdf

8   Bureau of Meteorology (BoM), 'New South Wales in summer 2016–17: warmest summer on record', Australian Bureau of Meteorology, 2017; bom. gov.au/climate/current/season/nsw/archive/201702.summary.shtml

9   Forbes, C., 'Inquiry into fire at "Flagview South", Sir Ivan Dougherty Drive, Leadville February 2017', Coroners Court NSW, Mudgee, 2019; coroners. nsw.gov.au/documents/findings/2019/Sir%20Ivan%20Doherty%20 Drive%20Leadville%20Fire%20Findings.pdf

10 Ibid.

11 Ibid.

12 CSIRO and BoM, *State of the Climate 2020*, 2020; bom.gov.au/state-of-the-climate/documents/State-of-the-Climate-2020.pdf

13 Keelty, M., 'Bega Valley Fires Independent Review', Office of Emergency Management, NSW Department of Justice, 2018; emergency.nsw.gov.au/ Documents/publications/Bega-Valley-Fire-Independent-Review.PDF

14 Bureau of Meteorology (BoM), 'Special Climate Statement 65 – persistent summer-like heat sets many April records', Australian Bureau of Meteorology, 2018; bom.gov.au/climate/current/statements/scs65.pdf

15 Ibid.

16 Ibid.

17 Burrows, N., 'Lessons and insights from significant bushfires in Australia and overseas: Informing the 2018 Queensland Bushfires Review', Bushfire and Natural Hazards CRC, 2019; igem.qld.gov.au/sites/default/files/ 2019-12/IGEM%20QBR%20BNHCRC%20-%20lessons%20and%20 insights.pdf

18 Ibid.

19 Inspector-General Emergency Management (IGEM), 'The 2018 Queensland Bushfires Review, Report 2: 2018–2019, Office of the Inspector-General Emergency Management, Queensland Government, 2019; https://www. igem.qld.gov.au/sites/default/files/2019-12/IGEM%20Queensland%20 Bushfire%20REVIEW%202019.pdf

20 Climate Council, 'Wet and wild: what to expect this summer', 2021; https://www.climatecouncil.org.au/resources/wet-and-wild-summer-2020-2021/

21 Dowdy, A.J. and Mills, G.A., 'Characteristics of lightning-attributed wildland fires in southeast Australia', *International Journal of Wildland Fire*, vol. 21(5), 2012, pp. 521–24; publish.csiro.au/wf/wf10145

22 AFAC, 'AFAC Independent Operational Review: A review of the management of the Tasmanian fires of December 2018 – March 2019', Australasian Fire and Emergency Service Authorities Council, South Melbourne, 2019

23 Littleproud, D., 'Doorstop with Mick Keelty at Parliament House, Canberra', 3 December 2019; minister.awe.gov.au/littleproud/speeches-and-transcripts/doorstop-keelty-mdbp

24 Jones, A., 'Testing truth by fire', *Daily Telegraph*, 19 November 2019

25 Binskin M., Bennett A. and Macintosh, A., 'Royal Commission into National Natural Disaster Arrangements – Report', Commonwealth of Australia, Canberra, 2020; naturaldisaster.royalcommission.gov.au/system/files/2020-11/Royal%20Commission%20into%20National%20Natural%20Disaster%20Arrangements%20-%20Report%20%20%5Baccessible%5D.pdf

26 Ciccarelli, R., 'NSW RFS Commissioner not warned over PM announcement army reservists deployed', 4 January, 2020; 9now.nine.com.au/today/nsw-bushfires-scott-morrison-didnt-tell-nsw-rfs-they-were-deploying-military-personnel/d21b5e42-ad12-47bd-88cf-62823802e946

27 Stayner, T., 'Defence Force chief says he was "discomforted" by Scott Morrison's bushfire response video', sbs.com.au, 4 March 2020; sbs.com.au/news/defence-force-chief-says-he-was-discomforted-by-scott-morrison-s-bushfire-response-video

28 Crowe, D., 'Deputy PM slams people raising climate change in relation to NSW bushfires', *Sydney Morning Herald*, 11 November 2019; smh.com.au/politics/federal/raving-inner-city-lunatics-michael-mccormack-dismisses-link-between-climate-change-and-bushfires-20191111-p539ap.html

29 Sullivan, K., 'National Farmers Federation calls for Australia to reduce net emissions to zero by 2050', abc.net.au, 20 August 2020; abc.net.au/news/2020-08-20/farmers-back-zero-emissions/12576806

30 Grattan, M., 'Grattan on Friday: When the firies call him out on climate change, Scott Morrison should listen', theconversation.com, 15 November 2019; theconversation.com/grattan-on-friday-when-the-firies-call-him-out-on-climate-change-scott-morrison-should-listen-127049

31 Culbertson, A., 'Australia PM Scott Morrison denies climate change link to bushfires', news.sky.com, 22 December 2019; news.sky.com/story/australia-pm-scott-morrison-denies-climate-change-link-to-bushfires-as-he-returns-from-holiday-11892929

32 Chang, C., 'PM Scott Morrison grilled on climate change in aftermath of devastating bushfires', news.com, 10 January, 2020; news.com.au/finance/work/leaders/prime-minister-scott-morrison-grilled-on-climate-change-in-aftermath-of-devastating-bushfires/news-story/323aac0df64b7493125350b95082198b

33 Park, A., 'Bushfire royal commission will fail if it does not recommend reducing emissions: Former fire chief', abc.net, 19 October 2020; abc.net.au/news/2020-10-19/bushfire-royal-commission-recommendations-climate-change-fears/12782484

34 Sanderson, B.M. and Fisher, R.A., 'A fiery wake-up call for climate science', pp. 175–77

35 Bureau of Meteorology (BoM), 'Special Climate Statement 71 – severe fire weather conditions in southeast Queensland and northeast New South Wales in September 2019', Australian Bureau of Meteorology, 2019; bom.gov.au/climate/current/statements/scs71.pdf

36 Queensland Reconstruction Authority (QRA), '2019 Queensland Bushfires: Recovery Plan 2019–2022', Queensland Reconstruction Authority, Brisbane, 2020; qra.qld.gov.au/sites/default/files/2020-08/2019_qld_bushfires_recplan_2019-20_lr_0.pdf

37 New South Wales Rural Fire Service (NSW RFS), *Bush Fire Bulletin*, vol. 42, no. 1/2020, 2020b, NSW Rural Fire Service, Granville

38 Ibid.

39 Binskin, M., Bennett, A. and Macintosh, A., 'Royal Commission into National Natural Disaster Arrangements – Report', Commonwealth of Australia, Canberra, 2020; naturaldisaster.royalcommission.gov.au/system/files/2020-11/Royal%20Commission%20into%20National%20Natural%20Disaster%20Arrangements%20-%20Report%20%20%5Baccessible%5D.pdf

40 *Guardian*, 30 December 2019, 'Volunteer firefighter Samuel McPaul died when "fire tornado" overturned 10-tonne truck'; theguardian.com/australia-news/2019/dec/31/volunteer-firefighter-samuel-mcpaul-died-when-fire-tornado-overturned-10-tonne-truck#:~:text=An%20emotional%20Rural%20Fire%20Service,old%20died%20at%20the%20scene.

41 Hannam, P., '"We're bloody lucky we didn't bury thousands of people": Constance's new climate pledge', *Sydney Morning Herald*, 12 October

2020; smh.com.au/environment/climate-change/we-re-bloody-lucky-we-didn-t-bury-thousands-of-people-constance-s-new-climate-pledge-20201009-p563lk.html

42 Arriagada, N.B., Palmer, A.J., Bowman, D., Morgan, G.G., Jalaludin, B.B. and Johnston, F.H., 'Unprecedented smoke-related health burden associated with the 2019–20 bushfires in Eastern Australia', *Medical Journal of Australia*, vol. 213 (6), 2020, pp. 282–3; mja.com.au/journal/2020/213/6/unprecedented-smoke-related-health-burden-associated-2019-20-bushfires-eastern

43 Boer, M.M., Resco de Dios, V. and Bradstock, R.A., 'Unprecedented burn area of Australian mega forest fires', *Nature Climate Change*, vol. 10, 2020, pp. 171–2; nature.com/articles/s41558-020-0716-1

44 The Australia Institute, 2020, 'Climate of the Nation 2020: Tracking Australia's attitudes towards climate change and energy'; australiainstitute. org.au/wp-content/uploads/2020/12/Climate-of-the-Nation-2020-cover-WEB.pdf

45 Readfern, G., 'Factcheck: Is there really a Greens conspiracy to stop bushfire hazard reduction?', *Guardian*, 12 November 2019; theguardian.com/australia-news/2019/nov/12/is-there-really-a-green-conspiracy-to-stop-bushfire-hazard-reduction

46 Di Virgilio, G., Evans, J.P., Clarke, H., Sharples, J.J., Hirsch, A. and Hart, M., 'Climate change significantly alters future wildfire mitigation opportunities in southeastern Australia', *Faculty of Science, Medicine and Health – Papers: Part B*, 2020; ro.uow.edu.au/smhpapers1/1513

47 CSIRO and BoM, *State of the Climate 2020*

48 Bureau of Meteorology (BoM), 'Special Climate Statement 72 – dangerous bushfire weather in spring 2019', Australian Bureau of Meteorology, 2019; bom.gov.au/climate/current/statements/scs72.pdf

49 RMIT ABC Fact Check Unit, 'Blaming arsonists? All the evidence points to a smokescreen', 2020; crikey.com.au/2020/01/16/arson-bushfire-crisis-fact-check/

50 Knaus, C., 'Bots and trolls spread false arson claims in Australian fires "disinformation campaign"', *Guardian*, 8 January 2020; theguardian.com/australia-news/2020/jan/08/twitter-bots-trolls-australian-bushfires-social-media-disinformation-campaign-false-claims

51 Owens, D. and O'Kane, M., 'Final report of the NSW Bushfire Inquiry', Parliament of NSW, Sydney, 2020

52 Owens, D. and O'Kane., M., 'Final report of the NSW Bushfire Inquiry'

53 RMIT ABC Fact Check Unit, 'Blaming arsonists? All the evidence points to a smokescreen'

54 Dowdy, A.J. and Mills, G.A., 'Characteristics of lightning-attributed wildland fires in southeast Australia', pp. 521–24

55 Department of Sustainability and Environment Victoria (DSE), 'Report of the investigation into the future of cattle grazing in the Alpine National Park', Alpine Grazing Taskforce, 2005; environment.gov.au/epbc/notices/assessments/victoria-alpine-national-park/pubs/b6-alpine-grazing-taskforce-2005.pdf

56 Ibid.

57 Luke, R.H. and McArthur, A.G., *Bushfires in Australia*, Australian Government Publishing Service, Canberra, 1978

58 Government of South Australia, 'Independent review into South Australia's 2019–20 bushfire season', Government of South Australia, Adelaide, 2020

59 Sanderson, B.M. and Fisher, R.A., 'A fiery wake-up call for climate science', pp. 175–77

60 Wang, B., Luo, X., Yang, Y-M., Sun, W., Cane, M.A., Cai, W., Yeh, S-W. and Liu, J., 21 October 2019, 'Historical change of El Niño properties sheds light on future changes of extreme El Niño', *Proceedings of the National Academy of Sciences of the United States of America*, 116 (45) 22512-22517; DOI: 10.1073/pnas.1911130116

61 Biskaborn, B.K., Smith, S.L. et al., 'Permafrost is warming at a global scale', *Nature Communications*, vol. 10, no. 264, 2019; nature.com/articles/s41467-018-08240-4

62 Lindsey, R., 'Climate change: atmospheric carbon dioxide', 2020; climate.gov/news-features/understanding-climate/climate-change-atmospheric-carbon-dioxide

63 Pörtner, H.-O., Roberts, D. C., Masson-Delmotte, V., Zhai, P., Tignor, M., Poloczanska, E., Mintenbeck, K., Nicolai, M. and Okem, A., 2019, 'IPCC special report on the ocean and cryosphere in a changing climate: summary for policymakers'; ipcc.ch/srocc/cite-report/

64 CSIRO and BoM, *State of the Climate 2020*

65 Abram, N.J., Henley, B.J., Sen Gupta, A. et al., 'Connections of climate change and variability to large and extreme forest fires in southeast Australia', *Communications Earth & Environment*, vol. 2 (8), 2021; nature.com/articles/s43247-020-00065-8

66 Arnell, N., Freeman, A., and Gazzard, R., 'The effect of climate change on indicators of fire danger in the UK', *Environmental Research Letters*, letter 16, 2021; iopscience.iop.org/article/10.1088/1748-9326/abd9f2

67 Luke, R.H. and McArthur, A.G., *Bushfires in Australia*

68 CSIRO and BoM, *State of the Climate 2020*, p. 4

69 Ibid.

70 Bureau of Meteorology (BoM), 'Special Climate Statement 71'

71 Abram, N.J., Henley, B.J., Sen Gupta, A. et al., 'Connections of climate change and variability to large and extreme forest fires in southeast Australia'

72 Ibid.

73 Bureau of Meteorology (BoM), 'Victoria in 2019: warmer and drier than average', Australian Bureau of Meteorology, 2019; bom.gov.au/climate/current/annual/vic/archive/2019.summary.shtml

74 Abram, N.J., Henley, B.J., Sen Gupta, A. et al., 'Connections of climate change and variability to large and extreme forest fires in southeast Australia'

75 Pendergrass, A.G., Meehl, G.A., Pulwarty, R. et al., 'Flash droughts present a new challenge for subseasonal-to-seasonal prediction', *Nature Climate Change*, vol. 10, 2020, pp. 191–9; nature.com/articles/s41558-020-0709-0

76 Hughes, L., Mullins, G., Rice, M. and Dean, A., 'Escalating Queensland bushfire threat: Interim conclusions', Climate Council, 2018; apo.org.au/sites/default/files/resource-files/2018-11/apo-nid208596.pdf

77 Hannam, P., 'Homes in Blue Mountains most at risk from bushfires, insurer says', *Sydney Morning Herald*, 22 August 2020; smh.com.au/environment/conservation/homes-in-blue-mountains-most-at-risk-from-bushfires-insurer-says-20200821-p55o0w.html

78 Luke, R.H. and McArthur, A.G., *Bushfires in Australia*

79 NSW Parliament Issues Backgrounder, 'Bushfires in NSW: Timelines and key sources', NSW Parliamentary Research Service, number 6, June 2014; parliament.nsw.gov.au/researchpapers/Documents/bushfires-in-nsw-timelines-and-key-sources/Bushfires%20in%20NSW%20-%20timelines%20and%20key%20sources.pdf

80 New South Wales Rural Fire Service (NSW RFS), 'Bush Fire Danger Period starts in a further 53 areas this weekend', NSW Rural Fire Service, 1 August 2019; rfs.nsw.gov.au/news-and-media/media-releases/bush-fire-danger-period-starts-in-a-further-53-areas-this-weekend

81 Luke, R.H. and McArthur, A.G., *Bushfires in Australia*

82 Bureau of Meteorology (BoM), 'Special Climate Statement 72'

83 Bureau of Meteorology (BoM), 'Special Climate Statement 73 – extreme heat and fire weather in December 2019 and January 2020', Australian Bureau of Meteorology, 2020; bom.gov.au/climate/current/statements/scs73.pdf

84 New South Wales Rural Fire Service (NSW RFS), *2019/20 Bush Fire Season Overview*, NSW Rural Fire Service, 19 March 2020, Granville

85 Bureau of Meteorology (BoM), 'Special Climate Statement 73'

86 Australian Academy of Science, 'The risks to Australia of a 3°C warmer world', 31 March 2021; science.org.au/news-and-events/news-and-media-releases/risks-australia-warmer-world

87 Fellner, C., 'Sky-high insurance threatens holidays', *Sun Herald*, 30 May 2021; parlinfo.aph.gov.au/parlInfo/search/display/display.w3p;query=Id%3A%22media%2Fpressclp%2F7985469%22;src1=sm1

88 Bowman, D., 'Prescribed burning: the pros, cons and unknowns,' Chapter 17 in *Prescribed Burning in Australasia: The science, practice and politics of burning in the bush*, Australasian Fire and Emergency Service Authorities Council Ltd., East Melbourne, 2020

89 Trask, S., 'Barilaro pours more fuel on bushfire spat', 7news.com.au, 13 November 2019; 7news.com.au/news/bushfires/barilaro-pours-more-fuel-on-bushfire-spat-c-555037

90 NSW Department of Planning, Industry & Environment, 'Accreditation of volunteer wildlife rescue and rehabilitation service providers in NSW', Parramatta, 2020; environment.nsw.gov.au/-/media/OEH/Corporate-Site/Documents/Animals-and-plants/Native-animals/wildlife-rescue-rehabilitation-service-providers-accreditation-190474.pdf

91 O'Sullivan, M., 'Homeowners allowed to rebuild after natural disaster using original planning rules', *Sydney Morning Herald*, 25 June, 2021; smh.com.au/national/nsw/homeowners-allowed-to-rebuild-after-natural-disasters-using-original-planning-rules-20210624-p583vb.html?btis

92 Snape, J., 'British Prime Minister Boris Johnson tells Morrison it's time for "bold action" on climate change', abc.net.au, 28 October 2020; abc.net.au/news/2020-10-28/boris-johnson-scott-morrison-climate-change-bold-action/12817474

93 The Australia Institute, 2020, 'Climate of the Nation 2020: Tracking Australia's attitudes towards climate change and energy'; australiainstitute.org.au/wp-content/uploads/2020/12/Climate-of-the-Nation-2020-cover-WEB.pdf

94 Hannam, P., '"Like terrorists": Malcolm Turnbull assails Liberal climate deniers', *Sydney Morning Herald*, 6 February 2021; smh.com.au/politics/federal/like-terrorists-malcolm-turnbull-assails-liberal-climate-deniers-20200206-p53y6u.html

95 Karp, P., 'Scott Morrison says no evidence links Australia's carbon emissions to bushfires', *Guardian*, 21 November 2019; theguardian.com/australia-news/2019/nov/21/scott-morrison-says-no-evidence-links-australias-carbon-emissions-to-bushfires

96 Snape, J., 'British Prime Minister Boris Johnson tells Morrison it's time for "bold action" on climate change'

97 Binskin M., Bennett A., and Macintosh, A., 'Royal commission into national natural disaster arrangements – report'

98 '"Zero emissions, zero jobs": Climate policy will "smash" regional industries, skynews.com.au, 20 April 2021; skynews.com.au/details/_6249422129001

99 Hannam, P., '"Like terrorists": Malcolm Turnbull assails Liberal climate deniers'

100 Luke, R.H. and McArthur, A.G., *Bushfires in Australia*

101 Owens, D. and O'Kane, M., 'Final report of the NSW Bushfire Inquiry', Parliament of NSW, Sydney, 2020

102 Gergis, J. and Carey, G., 'Some say we've seen bushfires worse than this before. But they're ignoring a few key facts', 14 January 2020; theconversation.com/some-say-weve-seen-bushfires-worse-than-thisbefore-but-theyre-ignoring-a-few-key-facts-129391

103 CSIRO and BoM, *State of the Climate 2020*

104 Bureau of Meteorology (BoM), 'Special Climate Statement 71'

105 Abram, N.J., Henley, B.J., Sen Gupta, A. et al., 'Connections of climate change and variability to large and extreme forest fires in southeast Australia'

106 Ibid.

107 Bureau of Meteorology (BoM), 'The air over Antarctica is suddenly getting warmer – here's what it means for Australia', Australian Bureau of Meteorology, 11 September 2019; media.bom.gov.au/social/blog/2195/the-air-above-antarctica-is-suddenly-getting-warmerheres-what-it-means-for-australia/#:~:text=means%20for%20Australia-,The%20air%20above%20Antarctica%20is%20suddenly%20getting%20warmer,what%20it%20means%20for%20Australia&text=Record%20warm%20temperatures%20above%20Antarctica,South%20Wales%20and%20southern%20Queensland.

108 Binskin, M., Bennett, A., and Macintosh, A., 'Royal commission into national natural disaster arrangements – report'

109 Abram, N.J., Henley, B.J., Sen Gupta, A. et al., 'Connections of climate change and variability to large and extreme forest fires in southeast Australia'

ACKNOWLEDGEMENTS

WHERE TO START? So many to thank: people who helped me, encouraged me, and gave me inspiration to write in the hope that the true impact of climate change could be communicated to a wider audience.

A good place to start is to thank my incredible parents, Pat and Jack Mullins. Although both have left us, their love, positivity and integrity continue to shape me and our extended family. Their strong social consciences and personal values of community service, respect and justice for all will always guide us. They taught me about the environment; bushfire fighting; the value of public service, of continuous education, facts, evidence and standing up for what is right. I am forever grateful that I had the good fortune to have them as my mum and dad, and I miss them.

My beautiful wife and best mate, Erris, has always inspired and amazed me. Caring, capable and calm, no matter what. While pursuing her own high-powered career and studies, she has always been my greatest supporter and confidant, and I try to be hers. Our children, Phil and Kate, and later our daughter-in-law, Andrea, all got used to me rushing off to fires in the midst of family gatherings, including Christmases. They just took it in their stride and seemed to understand that both their mum and dad were in caring professions, driven to help others. Our grandson, Eamon, always got a kick out of seeing 'Bop' head off with flashing lights and sirens, and our newest addition, Oli, also likes to see the 'wee-waa car' drive off when I'm on call, these days as a volunteer. I worry about what the future holds for their generation, and my love for them drives me to do as much as I can.

Growing up with my brother, Terry, and sisters, Kim and Robin, was fun, and yes, ladies, I'm so sorry that I burned down your cubby house when I was about five years old: truly, it *was* an accident. The three of us loved exploring the bush together, and Terry taught me how to snorkel, catch yabbies and to spearfish. Losing our big brother in 1972 scarred us, and I remember vividly the day we learned of his car crash. I realised years later that Terry's death was what drove me to pursue improved road accident rescue services.

On that note, I want to thank all of those who work on the front line as first responders: our firefighters (career and volunteer), police, paramedics, and SES volunteers, in whatever capacity you serve. You exemplify the ANZAC spirit and keep all of us safe, often at great personal peril and for little or no reward. You wear different uniforms and shoulder-patches, but you're all cut from the same cloth: heroes all. The general public have no idea of the danger, grief and fatigue that you endure on our behalf and the lasting toll that your service to others extracts from you and your families. I have been privileged and honoured to serve with the best of the best, here and overseas, and the emergency services will always be my second family.

To my colleagues in Emergency Leaders for Climate Action: each of you have had outstanding careers of service. You earned the right to peaceful retirements and had most certainly already contributed more than enough. Yet each of you readily stepped up to answer the call, despite the inevitable criticism and personal attacks when you had the courage to tell the truth about the true costs of climate change inaction. A number of you, like me, are now volunteers, and fought fires throughout Black Summer. We witnessed firsthand how drastically things have changed during our lives. Others were directly impacted by the fires and assisted with community recovery afterwards. Your efforts have inspired many, and I am deeply grateful to you for joining me on this crucial journey.

To the incredibly dedicated team at the Climate Council, there are not enough words. I have learned so much since being invited to join in 2018. The diverse backgrounds and knowledge of the Council and Board members, the amazing staff and volunteers, have enabled the Council to achieve so much. Particular thanks to Lisa, Martin, Amanda, Tim, Lesley, Will, Katrina, Alix, Vai, Violette, Morgan, Anni, Simon and all the others who have helped me over these past few years, contributing their knowledge, passion and hard work to ensure that we get the climate message out there, particularly in trying to make it okay for people on the conservative side of politics to acknowledge science and facts, and that we have limited time left to act.

A big thanks to my agent, Jane Novak, to Nikki Christer and Clive Hebard at Penguin Random House, and to Liam Pieper for his help with an early draft. Nikki and Jane in particular believed in me from the start, and this book wouldn't have happened without them.

Lastly, to the readers. Thank you for reading this book, and for letting me explain to you why people like me who have fought fires and natural disasters for decades are . . . scared. Our world is rapidly changing, and we don't have a great deal of time left to take the necessary actions that might still give us a chance to limit the damage. I hope that you are moved to take responsible action and to demand more from our national government, before it is too late: our kids and grandkids are worth the effort.

Greg Mullins AO AFSM became a major national figure in the 2019–20 bushfire crisis – Australia's longest, hottest and most devastating on record. From being a volunteer firefighter and a career firefighter, he became an internationally recognised expert in responding to major bushfires and natural disasters. During his thirty-nine-year career, culminating in being Commissioner of one of the world's largest fire services for nearly fourteen years, he served as President, Vice President and Board Chair of the Australasian Fire & Emergency Service Authorities' Council, Deputy Chair of the NSW State Emergency Management Committee, Australian Director of the International Fire Chiefs Association of Asia, NSW representative on the Australian Emergency Management Committee, member of the NSW Government Climate Change Council, Australian representative on the UN's International Search and Rescue Advisory Group, and as a member of the NSW Bushfire Coordinating Committee. He is Chair of the NSW Ambulance Service Advisory Board and a Climate Councillor. In early 2019, fearing an imminent bushfire disaster, he formed Emergency Leaders for Climate Action, a coalition of thirty-four former fire and emergency service chiefs from throughout Australia. ELCA continues to call for action at all levels of government on climate change and greenhouse gas emissions, explaining how climate change is super-charging extreme weather, bushfires and other natural disasters.